Radio Controlled

MODEL AIRPLANE

BASIC FLIGHT TRAINING

COURSE MANUAL and GUIDE

Philip E. Balcomb

Radio Controlled Model Airplane

BASIC FLIGHT TRAINING
COURSE MANUAL and GUIDE

Copyright © 1992 by Philip E. Balcomb

Reproduction by any means without written permission of the Copyright owner is against the law.

ISBN 0-9620456-2-4

Published by

TEMPUS Press
P.O. Box 235, Tell City, IN 47586

Printed in the United States of America

Radio Controlled Model Airplane

BASIC FLIGHT TRAINING

COURSE GUIDE

INTRODUCTION

This course is designed for use by instructors who are accomplished pilots but who may have had little or no experience in effective educational techniques. It is intended to be conducted in relatively short sessions, ideally with one, but not more than three students at a time, at a pace that is comfortable for each student. It is inevitable that some students will progress more rapidly than others. The instructor, before proceeding from one step to the next should question each student on the material just covered to make sure that he or she understands it and all of his questions on that subject have been answered.

Theoretical aspects of flight have been conscientiously avoided. Only when a student has become at least an intermediate flyer will the refinements brought out by theory be of value to him or her. This is based on the evident fact that, with sufficient power and control surfaces, almost any shape can be made to fly. Witness the flying brooms, irons and ironing boards, washboards and other novelties that appear at major model fly-ins.

Our objective is to simply show a student, through logical steps, how he or she can comfortably and safely start, taxi, take off, fly and land a forgiving trainer airplane. After gaining experience in these elements, there will be time to consider more advanced aircraft and forms of flying.

A copy of this Guide should be provided to each student about a week before the first scheduled training session. A trainer aircraft in ready to fly condition and its radio transmitter should be available at each session for hands-on demonstration.

The course is divided into two parts, Pre-Flight Training and Flight Training. Only after completing the Pre-Flight portion should the student be introduced to the Flight Training part.

It is strongly recommended that all but Field and Flight training be conducted indoors in a location having a table large enough to support a Trainer Plane and Radio Transmitter with comfortable chairs for instructor and students.

This COURSE GUIDE should be followed step by step, with hands on demonstration, and plenty of time for the asking and answering of questions and discussion of any pertinent aspect of R/C flying.

Thorough understanding of the basic fundamentals contained in the first sections of this Guide will make the actual process of learning to fly much more rewarding for both student and instructor.

TABLE of CONTENTS

STUDENT LOGS vi

PART I   Pre-Flight Training 1
          Ground School

PART II Pre-Flighting Your Plane 35
          Field Equipment and Procedures

PART III Primary Flight Training 65

Typical AMA Club Safety Rules 79

INDEX 85

STUDENT TRAINING LOG 1

Student: _____ Instructor: _____

Date	Subject	Passed
	Identification and Function of Plane Elements	
	Identification and Function of Engine and Related Accessories	
	Wing Attachment	
	Identification and Function of Radio System Elements; Receiver, Transmitter, Servos, Batteries, Chargers, Antenna, Switches, Channels, Safety Procedures	
	Operation of Control Sticks, Receiver, Servos, Push Rods, Control Horns, Clevises, Adjusting Control Surface Movement, Trimming, Combined Control	
	What Makes Planes Fly, Wing Lift, Effect of Air Currents and Wind	
	Directional Control on the Ground, Taxiing under Control	
	The Third Dimension, Moving up and Down in the Air	
	Introduction to R/C Flying, Importance of Visual Contact	

STUDENT TRAINING LOG 2

Student: _____   Instructor: _____

Date	Subject	Passed
	Understanding Flying Field Operation and Safety Procedures	
	Familiarity with Field Equipment; Field Box, Fueling, Electric Starter, 12 Volt Battery and Charging, Fuel Pumps, Glow Plug Heaters, Trickle Chargers, Tachometers.	
	Radio Procedures; Frequencies, Channel Identification, Control Systems	
	Pre-Flight Procedures; Wing Attachment, Fueling, Structural Check, Receiver Battery Check, Reserving Radio Channel, Turning on Reciever and Transmitter, Radio Range Check, (Signals).	
	Engine Starting and Adjusting; Priming, Glow Plug Heater, Idle Setting, Attitude	
	Taxiing; Race Track Pattern, Figure Eight, Between Markers, Through Gates	
	Shutting Down; Radios OFF, Plane Cleaned, Fuel Drained, Site Left Clean	

STUDENT FLIGHT TRAINING LOG 3

Student: _____   Instructor: _____

Date	Subject	Passed
	Taking Control in the Air; Coordinated Turns, Maintain Altitude, Fly Pattern	
	Right and Left Turns, Figure Eights, Change Altitude and Maintain Pattern, Vary Throttle Settings from Half to Full	
	Reverse Pattern; Climbing and Descending Turns, Flyby Simulation of Landing Pattern	
	UNASSISTED TAKEOFF; Repeated Practice, Demonstrate Ability to Maintain Direction During Takeoff and Climb, Climbing Turn into Pattern	
	LANDING; Practice Low Level Simulation, Touch and Go, Landing	
	Recovery from Simulated Out of Control Situations at Altitude	
	Practice Takeoff and Landing, with Instructor Present	
	CAPABLE of FLYING SOLO Safely, Without an Instructor Present	

PART I

PRE-FLIGHT TRAINING

GETTING ACQUAINTED with YOUR MODEL

Elements of a Radio Controlled Model Airplane

The WING

The wing provides the lift that enables the plane to overcome the pull of gravity. It is usually made up of light weight wood elements assembled to form a frame. Each half of this frame is usually constructed as a unit, then the two are joined at their inner ends. After assembly, the wing is covered with a heat-shrinkable plastic film.

RIBS

Trainer Wing ribs are flat on the bottom, are flat at the front to accept a rounded nose (*leading edge*) and rise slightly to a point about a third of the way back, then taper to almost a point at the back where a spar (*trailing edge*) is attached. Each rib is notched to accept the spars.

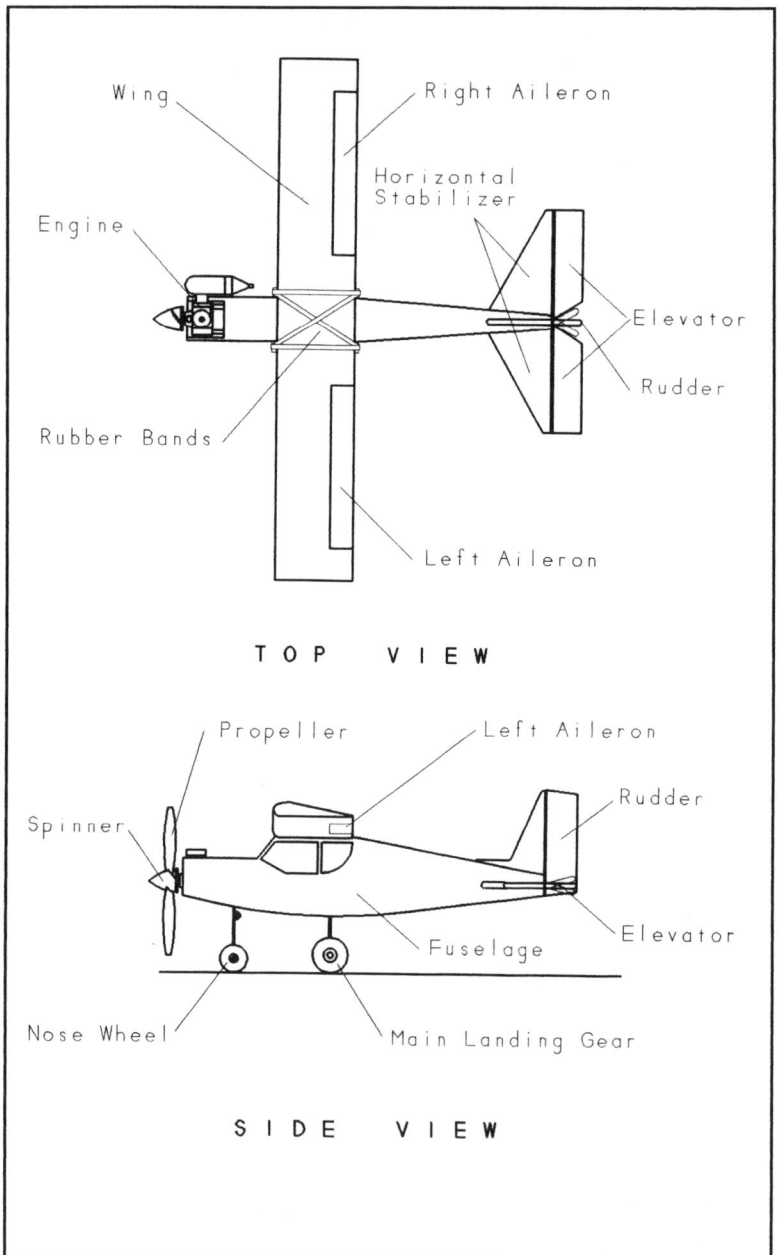

Figure 1 Elements of Your Airplane

SPARS

These are long sticks that form the backbone of the wing. There are usually two main spars, plus a leading edge and trailing edge piece. The ribs are spaced evenly along the spars. Notches in the ribs fit snugly over the spars and are glued in place. The leading edge spar may be rectangular, to be rounded off after assembly, or may be a round dowel that fits into a semi-circular cutout in the nose of the ribs. The trailing edge spar may be rectangular, tapered later to follow the line of the top of the rib, or may be preshaped.

AILERONS

Not all trainer wings have ailerons, but most do. An aileron is a section of the outer trailing edge that is separated from, but hinged to the wing so that it can be moved up and down. They are *control surfaces* used to control the plane in turns.

Control Horn

Attached to each aileron, at right angles to its surface, is a tab with tiny holes into which the end of a control cable or rod can be inserted to move the aileron up or down.

Radio Controlled SERVO

In most trainer planes, a servo is mounted on the under side of the wing to operate the ailerons. A servo is a tiny electric motor driving a gear train than greatly reduces the output speed. On the top of the servo is a plastic arm or wheel with holes into which a clevis at the end of control rod or cable is attached. The motor is activated by a signal from the Radio Receiver carried on board the plane. The arm will move in direct proportion to the movement of the control lever on the Radio Transmitter. This will be demonstrated in due course.

Wing Covering - Monocote or similar material

There are a number of covering materials, the most commonly used of which is Monocote. This is a very thin, very tough, colored plastic film. It has a heat sensitive adhesive backing on one side. When the material is stretched over the wing framework, the tip of a heated iron is used to attach it to the ribs, spars, leading and trailing edges. After the film is firmly secured, the flat of the iron is passed over the surface, causing it to shrink and become very taut.

The FUSELAGE

This is the body of the aircraft to which are attached the Wing, the Empenage or "Feathers" (Tail Assembly) and Wheels. The fuselage carries the Engine, Fuel Tank, Radio Receiver, Batteries, Servos and Control Rods.

Construction

Most Trainer Plane Fuselages are made up of a combination of sheet balsa and plywood and balsa stringers. Covering is the same as the wing.

Tail Assembly

Stabilizers

The *Vertical* Stabilizer, sometimes called a 'fin' is attached

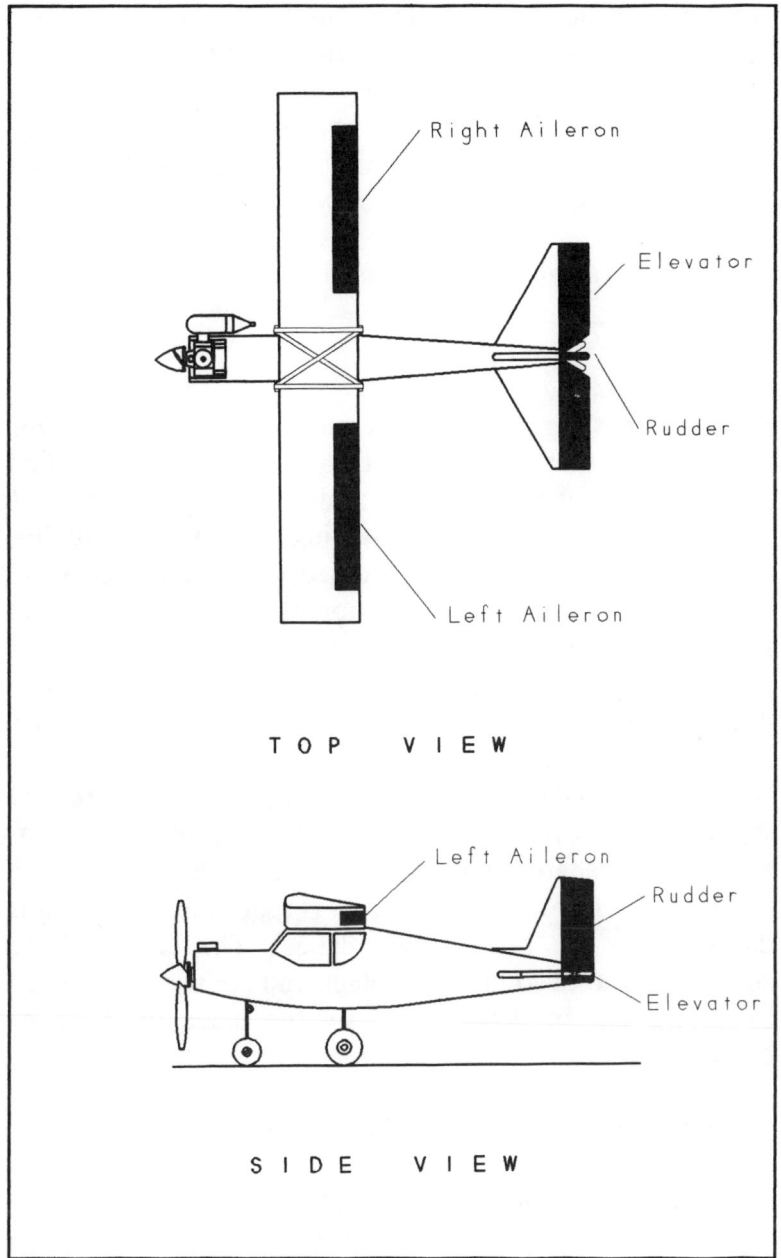

Figure 2 Control Surfaces

to the top of the rear end of the fuselage. The *Horizontal* stabilizers are attached, one to each side of the rear of fuselage. Stabilizers may be made up of strips of balsa, or may be a solid sheet of balsa. They are covered like the wing. Stabilizers act like the tail feathers of an arrow to help keep the plane flying straight and true.

Control Surfaces

Rudder

The Rudder is attached to the rear of the Vertical Stabilizer by hinges, so that it can move from side to side. Attached to it is a Control Horn, like the ones on Ailerons. It is of the same construction and covering as the Vertical Stabilizer. The rudder is moved by a servo in the fuselage and is used to help control the direction of travel on the ground and, to a lesser extent, in the air.

Elevators

An elevator is attached at the rear of each of the Horizontal stabilizers with hinges, so they can move up and down. Usually, they are connected to each other and move together, so only one control horn is required. Covering is the same as the rest of the plane. The elevators are activated by a servo in the fuselage and are used to raise or lower the tail in flight, to cause the plane to climb or descend.

The ENGINE, Fuel Tank and Propeller

The single cylinder internal combustion engines used to power R/C Trainer Planes are simple but very sophisticated mechanisms. They are nearly always what is known as *two*

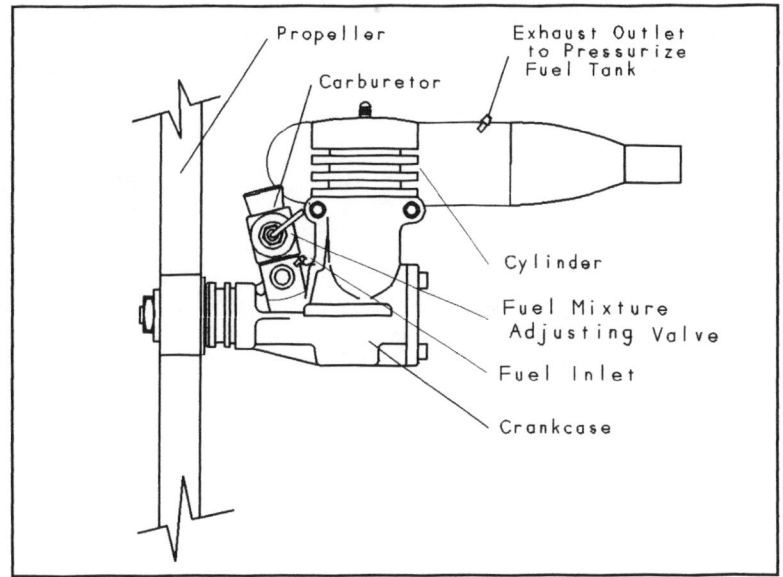

Figure 3 Typical R/C Engine - Side View

stroke, that is, there are two strokes in each combustion cycle. These engines burn alcohol to which castor or synthetic oils are added. Some fuels contain from 5% to 25% nitro methane for additional power.

Engine

R/C engines consist of a crankcase, cylinder, crankshaft, piston and connecting rod, carburetor and muffler. The two cycle engines used in nearly all trainers have no valves or ignition system. These engines run at very high speed and those used in trainers produce almost a full horsepower. This power must be respected.

Glow Plug

While in gasoline powered engines a spark plug and electri-

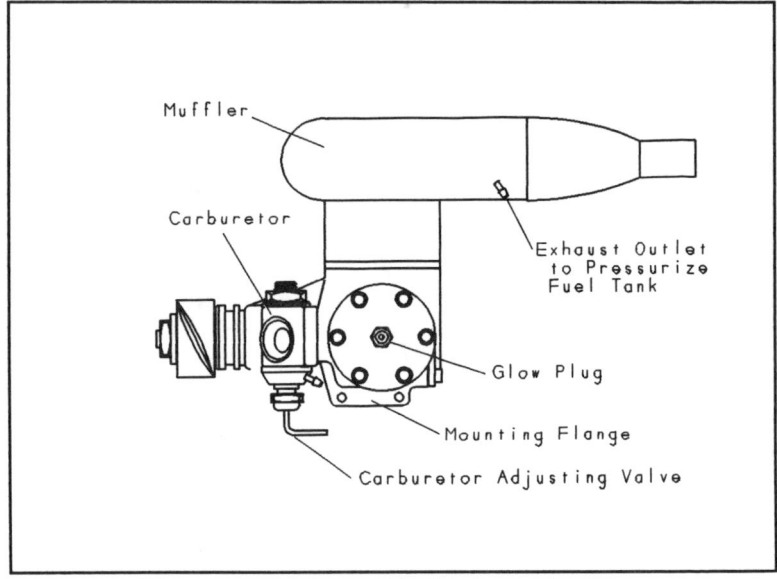

Figure 4 Typical R/C Engine - Top View

cal ignition system are required, your R/C engine uses a *glow plug*. Glow plugs look much like very small spark plugs. They have a threaded body and a knobbed stem sticking out of the top. They are screwed into the cylinder head of the engine, like a spark plug. The business end, however, instead of having two electrical conductors with a gap between them to produce a spark, has a tiny tungsten wire attached to the center conductor and to the body. When a current is passed through this wire, it becomes red hot.

To start the engine, a battery powered connector is snapped over the end of the glow plug. This creates an electrical circuit between the center contact and body, causing current to flow through the tungsten wire, raising it to red heat.

When the crankshaft of the engine is rotated the fuel-air mixture is compressed and the heat of the wire ignites it. The heat of repeated explosions of the gases keeps the wire hot, so that after each cycle, it is hot and ready to ignite the next fuel charge.

Engine Mounting

Engines in Trainer planes may be mounted on two beams that extend out from the first bulkhead or former, called the *firewall* and secured with four recessed head screws. In some cases, a special mount is attached to the back of the engine and to the firewall.

Fuel Tank

A plastic tank, usually with a capacity of about four ounces is installed just behind the engine. A silicone plastic tube with a heavy brass fitting called a *clunker* at its end is inserted inside the tank and the other end connected to the intake of the engine. A second tube runs from the engine muffler to the tank, to create a little pressure to help force fuel to the engine. The tank is usually mounted in a compartment and is cushioned and held in place by plastic foam pads.

Throttle Control

The carburetor is mounted on the engine's front side. A lever with several small holes projects from the right side of the carburetor and can be moved forward and backward. This movement opens or closes the carburetor throat and controls the amount of fuel and air ingested into the engine, controlling its speed. The lever is attached by a control cable to a servo mounted further back in the fuselage.

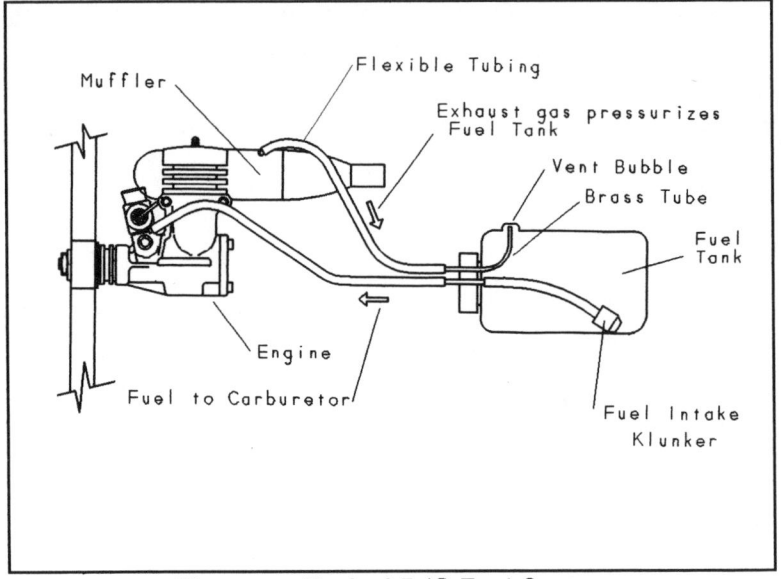

Figure 5 Typical R/C Fuel System

On-Board Fuel System

A plastic fuel tank is stowed in the plane fuselage. It usually is rectangular in shape and holds about 4 ounces of fuel. Two small brass tubes pass through a rubber stopper that seals the tank. On the inside of the tank, one tube is bent up so that its end is inside a small bubble in the top of the tank. A silicone fuel line is attached to the outer end a silicone tube is attached to this brass tube. Its other end is then connected to a fitting on the muffler. Back pressure from the exhaust keeps pressure on the tank to help force fuel to the engine.

To the inside of the second brass tube a short length of silicone tubing is connected. A short brass cylinder with a hole in it called a *clunker* is attached at the other end. It is heavy and the weight of the clunker keeps it in the fuel

when the plane is on its side or upside down. If the fuel supply is low, and the plane is in a steep dive, the fuel will flow to the front of the tank and the clunker may remain above the fuel.
Propeller

The Propeller

The propeller may be made of wood or of one of several plastic compounds. For efficiency, its edges are usually very sharp. When spinning at many thousands of revolutions per minute, they act as knives and can inflict severe injury to hands and fingers. The tips should be painted a bright color to increase visibility.

Landing Gear

Most Trainer Planes have a tricycle landing gear. Two main wheels are mounted on heavy wire or aluminum struts firmly attached to the bottom of the fuselage. A third wheel is mounted on another heavy wire strut that runs through and is supported by blocks attached to the fire wall. It is coiled near its middle to provide impact resistance and near the top is attached a steering lever. This lever is connected to the same servo that controls the rudder, enabling steering of the nose wheel for ground maneuvering.

Wing Attachment

Most wings of Trainer Planes are attached to the fuselage by heavy rubber bands. Two dowels run through the fuselage and extend out about ½" on each side. One is located just ahead of the intended position of the leading edge and the other just behind the trailing edge of the wing.

Several rubber bands are stretched from one dowel to the other, over the wing on each side. Sometimes a few are crossed diagonally over the wing.

The RADIO SYSTEM

The Receiver

The Radio Receiver, while quite small, is highly sophisticated. Its antenna, a flexible wire about three feet long, is permanently attached. The receiver has five to eight tiny sockets into which servo plugs and battery plugs are inserted. The receiver is usually mounted as far forward as possible, is cushioned with foam and wrapped in plastic film, secured with masking tape.

The Batteries

To power the Radio Receiver, four AA Size rechargeable batteries are shrink-wrapped together to form a pack. A wire with a plug is attached for connecting it to the Radio Receiver. This is a fairly heavy unit and is usually stowed as far forward as possible in the fuselage. It is cushioned by wrapping with plastic foam and often protected with plastic wrapping film.

The Charge Connector and OFF-ON Switch

A wire harness with suitable plugs and an ON-OFF switch is connected to the Radio Receiver and the Battery Pack. This harness also has a receptacle plug into which the charger plug can be inserted to recharge the batteries. The Switch Receptacle is mounted at a convenient location on the fuselage. The charge receptacle may be left loose inside the fuselage, or, with an adapter fitting, may be mounted so that it is accessible from outside the plane. The switch should be kept in the *OFF* position at all times when the radio is not in use, to avoid battery drain.

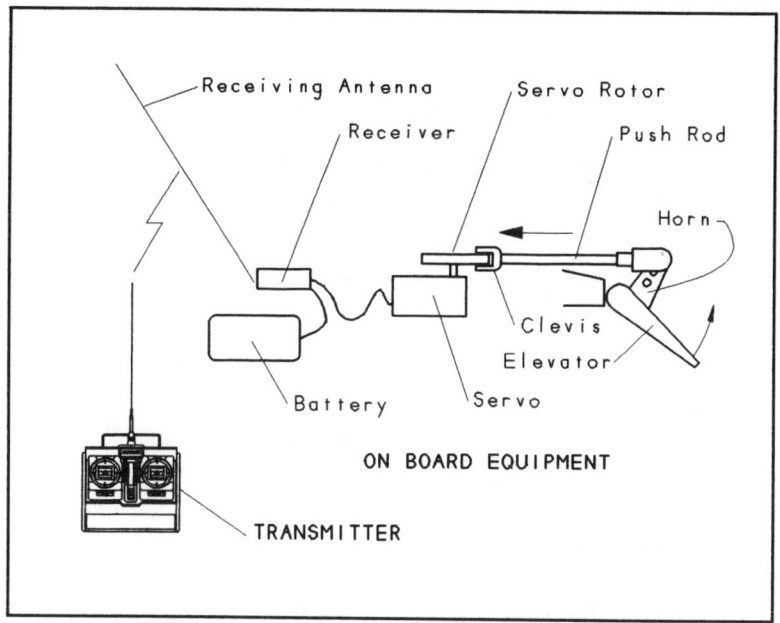

Figure 6 Typical R/C Radio System

The Servos

The servos for the Elevator, Rudder and Throttle are usually mounted on a plywood sheet or other mounting inside the fuselage, in the uncovered area under the wing. The servo for the Throttle is usually mounted transversely at the front, the other two side by side just behind it. A cable running through the fire wall connects the servo to the throttle. Usually rods with threaded wires attached at each end for clevises connect the other servos to the elevators and to the rudder and nose wheel.

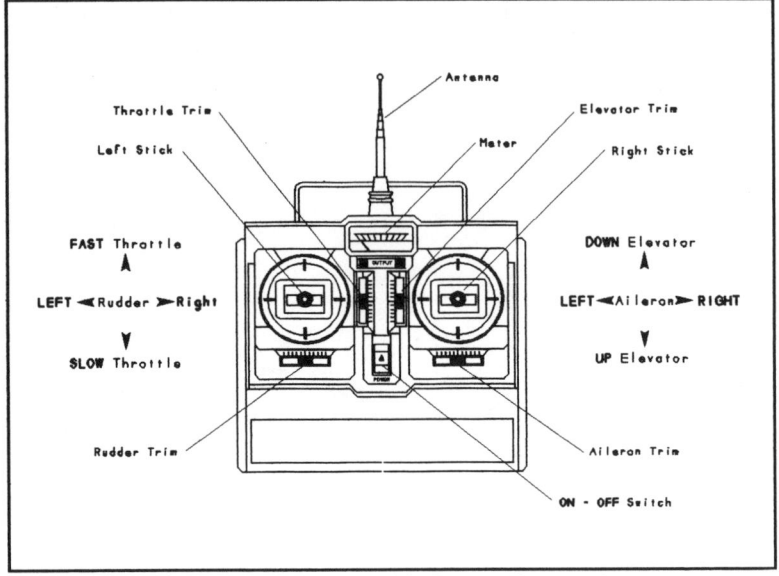

Figure 7 Typical R/C Radio Transmitter

The RADIO TRANSMITTER

The Transmitter that sends radio signals to the Receiver in the plane is mounted in a case that also contains the antenna, a rechargeable battery pack, with a charge receptacle, a meter showing battery condition and the levers and slide switches necessary to effectively control the plane. In addition there may be other special controls and accessories.

ON-OFF Switch

This is usually a slide switch, with one end marked ON and the other OFF. When the switch is in the ON position, current flows from the batteries to the radio. If left on, the batteries will be drained and possibly damaged.

*ALWAYS KEEP THE TRANSMITTER OFF-ON SWITCH IN THE **OFF** POSITION WHEN NOT ACTUALLY USING THE RADIO TRANSMITTER.*

Telescoping Antenna

The antenna can be collapsed inside the transmitter case, so that it is easier to transport and store.

WHEN THE TRANSMITTER IS IN USE, the ANTENNA MUST BE FULLY EXTENDED

Failure to extend the antenna will result in limited or erratic signal transmission and loss of control of the plane.

Charging Jack

Usually located on the side of the transmitter case is a female jack, much like that for earphones on an ordinary radio. Supplied with your radio is a battery charger that plugs into a wall outlet. A plug attached to the charger fits the jack on the transmitter. A light on the charger glows when the battery is connected.

Follow charging instructions shipped with your radio.

OVERCHARGING can SERIOUSLY DAMAGE BATTERIES.

Control Sticks

Centered on each side of the front of the transmitter case is a control lever. These levers pivot around their ends, inside the transmitter, and can be moved in all directions.

Left Control Lever (Stick)

The control lever on the left is used to control the *rudder and nose wheel*; as it is moved **right** or **left**, the rudder and nose wheel respond with similar motions. When released, the lever returns to center, as do the rudder and nose wheel.

The left lever also controls the *throttle*. Moving the lever **up** increases the carburetor opening, while moving it **down** reduces it.

Unlike the horizontal movement of the control lever associated with the rudder and nose wheel, the throttle control is *not spring loaded* and the lever remains in the position it is in when released. Engine speed continues at the set level.

Right Control Lever

The right lever on the transmitter is spring loaded in all directions. When it is released, it will return to the true center position.

Moving the lever **up** and **down** controls the movement of the *elevator*. Moving the lever from **side to side** controls the motion of the *ailerons*. Moving it to the *left* causes the *left aileron* to go **up** and the *right aileron* to go **down**. Moving the lever to the *right* reverses the action. When the lever is centered, the top and bottom surfaces of the aileron are flush with the wing surfaces.

Combined Aileron - Elevator Control

If the lever is moved at an angle between vertical and horizontal, *both the elevator and the ailerons are affected*.

Slide Controls for TRIM

Ideally, when both control levers are centered, the plane should fly straight and level at medium speed. Before the first flight, an experienced pilot will help you set up your plane so that all flight surfaces are centered when the controls are centered and the throttle is properly positioned.

Even though everything appears right, however, in flight, minor adjustment is usually required to achieve neutral straight and level flight. This is called *trimming*. To accomplish this, very fine adjustments are needed and can be made with four sliding trim controls. These controls are mounted alongside or under the control levers and they move in the same direction as the levers to which they relate. They are not spring loaded and stay in the position they are left in.

Throttle Trim is usually accomplished on the ground, using the slide control just to the right of the left control lever.

Rudder and Nose Wheel Trim

The trim slide control for left and right movement of the rudder and the nose wheel is mounted under the left control lever and moves to left and right, paralleling the motion of the lever.

Elevator Trim

The elevator trim slide control is located to the left of the right control lever. It moves up and down, as does the lever as it controls elevator movement.

Aileron Trim

The aileron trim slide control is located under the right control lever, paralleling the motion of the lever as it actuates the ailerons.

Servo Reversing Switches

It is essential that each servo move in the right direction. If one or more do not, their action can be reversed by changing small slide switches usually mounted on the bottom of the transmitter case. Your instructor will help you make sure these switches are properly set. After that, they will be of no concern.

HOW the RADIO CONTROLS OPERATE

The identifying plate on your transmitter, in addition to its trade name and other information, will probably include the phrase, *Digital Proportional Radio Control System*. This simply means that the servos driving the movement of the various control elements move in direct *proportion* to the movement of the control levers.

For example; if the throttle control stick is moved halfway up, the carburetor throat will be opened halfway. Moving the lever all the way to the top fully opens the throttle.

The full travel of a servo arm is 90°, divided into 512 segments. If a control lever is moved from neutral to one end of its travel, 256 pulses will be sent to the receiver and the servo motor will be activated 256 times to move the arm 45°. For shorter travel, fewer impulses are sent. Each impulse moves the servo arm a tiny amount, hardly visible to the human eye.

If the stick is moved a short distance, only a few pulses are sent. Each pulse moves the servo arm one segment and it stops at the same relative position as the lever. When the stick is released and the spring returns it to neutral, an equal number of reverse pulses are transmitted and the servo returns to neutral.

The servo responds not only to the *distance* its lever moves, but also to the *speed* at which it moves. Rapid motion of the control lever results in rapid motion of the servo and of the element it operates. Normally, it is best not to move the control sticks fast, or with jerky motions. Smooth control will result in smooth maneuvers.

Thumb Only, or Thumb and Forefinger Control?

The outer ends of the control sticks are rough to the touch. This is so that the thumb, when placed on it, will not easily slide off. Many R/C pilots use only the thumbs for moving the sticks. They reason that the thumb is the most maneuverable of all the fingers and another finger only gets in the way, while requiring the pilot to be dexterous with both.

Other good pilots feel more comfortable with the thumb and forefinger, manipulating the lever like a pencil. You should try both methods before you get into the air, to see which you prefer. Choose the one that is most comfortable for you. You can change your mind later.

Light Touch

Whichever system you decide on, only very light pressure be exerted on the sticks. You should be able to feel the resistance of the centering springs and to detect when the stick is in the neutral position.

Practicing Operating the Controls

With your plane an a table in front of you and your radio with fully charged batteries, move the receiver switch on the plane to the ON position. Do the same with the switch on the transmitter. Your transmitter is now sending out a carrier wave that is being received by the receiver in the plane.

It is probable that you heard a slight *whirrr* when you turned the switches on. If the servos in the plane were not in exactly the position required by the levers on the transmitter, the signals sent from the transmitter cause them to move to the proper position.

Throttle Control

Move the left control stick slowly up and down, stopping anywhere you like. Observe the action of the throttle control arm on the carburetor. Move the stick very slowly. You will see that the throttle arm also moves very slowly.

Move the stick rapidly and notice that the arm moves fast. Notice that whenever and wherever you stop moving the stick, the throttle arm stops, too. Note, also, that, unlike other controls, the throttle stick does not return to center.

Throttle Trim Control

The ideal setting for the throttle when the stick is all the way down produces a slow idling of the engine. If you look into the throat of the carburetor, you will see a very small opening. Your instructor will make adjustments in the control linkage to obtain this condition. The Throttle Trim slide control to the right of the stick should be all the way to the top. With this setting, the engine can be comfortably idled at any time.

To shut the engine down completely, the trim control is simply moved to the bottom of its slot, fully closing the carburetor throat. Observe this action several times to be sure you understand how it works. Don't hesitate to ask your instructor about anything you don't fully understand.

Rudder - Nose Gear, Avoiding Unintentional Movement

Moving the left stick to left or right causes a corresponding movement of the servo controlling the rudder and nose wheel. Try moving the stick fast, then slow and observe that the action of the wheel and rudder correspond. When you release the stick, it automatically returns to the center.

Since the left stick controls both the throttle and the rudder, it is important that you be able to move the stick to change one, without disturbing the other. Practice moving the stick up and down to open and close the throttle, with moving the rudder. Also practice changing the throttle setting without moving the rudder.

Rudder Trim Control

The slide trim control for the rudder is located underneath the left stick. Before the first flight your instructor will adjust the linkages so that the rudder is centered when both the stick and the trim control are centered.

Try moving the trim to one side or the other and observe the very small motion resulting in the rudder and nose wheel. This makes possible precision tuning for maximum control.

Elevator Control

Up and down motion of the right stick moves the elevator up and down. Notice that, if you move the stick even slightly to the side as you go up and down, the ailerons are affected. Again note that the elevator responds directly to the movement of the stick, fast or slow. When pressure on the stick is released, the elevator returns to center.

Elevator Trim Control

The elevator trim control is just to the left of the right stick. Your instructor will adjust linkage so that the elevator is centered when both the stick and the trim control are centered. Move the trim up and down and notice the resulting movement of the elevator.

Aileron Control

Left and right movement of the right stick controls the movement of the ailerons. Note that the left and right ailerons move in opposite directions. Moving the stick to the left causes the left aileron to go down and the right one to go up. Moving the stick to the right reverses this action. Again note that if you move the stick fast the response of the ailerons will be fast. If you move it slowly, the ailerons move slowly. Note also that when pressure on the stick is released it returns to center, as do the ailerons.

Aileron Trim Control

The aileron trim control is located below the right stick. Your instructor will have adjusted the linkage so that the ailerons will be centered when both the stick and the trim control are centered.

Combined Control - Aileron and Elevator

The right control stick can be moved in any direction, between vertical and horizontal. This means that if it is moved slightly up and to the left, the elevator will go down at the same time the left aileron moves up and the right aileron down.

Try moving the stick in different directions and watch the action of both elevator and ailerons.

Most turns in the air require this combination of control. When properly synchronized, it will produce coordinated turns, without losing altitude.

It is important to be able to feel the position of the stick, so you know what the control surfaces of your plane are doing when you cannot actually see them. Practice moving the stick up and down so that *only the elevator* moves, with no motion of the ailerons.

Now move the stick from side to side, so that *only the ailerons* move, without any motion of the elevator. It may take some time to master this, but it will be worthwhile.

SHUTTING DOWN the RADIOS

As soon as you are through practicing with the radio, move the switch on the plane to the **OFF** position. Move the switch on the radio transmitter to the **OFF** position. Follow the instructions of the manufacturer with regard to recharging batteries.

Overcharging or charging before the batteries are discharged to an appropriate level may result in damage to the batteries. On the other hand, a sufficient charge should be retained to allow some flying. Your instructor can tell you more about the care of batteries.

SAFETY - The Most Important Consideration

R/C Models are not toys. They are serious flying machines that must be treated with respect. They operate at speeds that make them potentially dangerous missiles, unless competently controlled. The propeller, driven by an engine that produces significant horsepower, spins at very high speed and its edges are sharp enough to cause serious injury. Both on the ground and in the air, SAFETY must have first priority.

AMA and Club SAFETY CODES (See back of this book)

Every flyer of Radio Controlled Model Aircraft should be a member of the *Academy of Model Aeronautics* (AMA). Nearly all Radio Controlled Aircraft Clubs require membership in AMA as a condition of membership in their club. When applying for membership, the applicant must sign a statement that he will comply with AMA Safety Rules. These rules are supplied to each member. They are solely for the benefit of the member, to help him avoid accidents. Every student flyer should be familiar with all these rules and should ask the instructor about any areas that are not clear.

WHAT MAKES AIRPLANES FLY ?

An only slightly oversimplified answer to this question is *POWER*. We have all watched the television screen in awe as a space vehicle weighing hundreds of thousands of pounds is forced into the air by the sheer power of its rocket engines that force it vertically into the air and build its speed to over 17,000 miles an hour to put it into orbit around the earth.

A simpler demonstration of the truth of this statement is in the form of strange objects, such as brooms, doghouses and other non-aerodynamic shapes that are flown, under control, by innovative R/C modelers. They fly only because they have sufficient POWER. Their performance capabilities are very limited and it is probable their owners breathe a sigh of relief if they are landed safely.

If you have ever put your hand into the air through which a car is moving at some speed, you found that a fist was simply pushed back and required little force to keep it in place. When the hand is opened up to create a larger surface area, much more energy is necessary to hold it vertical. On the other hand, if the fingers are pointed directly into the air stream, the resistance is very substantially reduced. Raising the tips of the fingers slightly pushes the hand up. Lowering the tips of the fingers pushes the hand down. The position of your hand, above or below the horizontal is its *attitude*. If the car slows down, the pressure on your hand is reduced by the slower flow of air until there is no perceptible force exerted on it at all and your arm tends to fall, having lost its LIFT.

The WING Provides LIFT - Providing it is MOVING.

The wing of an airplane, regardless of its shape, behaves much the same as your hand. When the engine and propeller pull it through the air at sufficient speed and its attitude is slightly up, it tends to rise. This would occur even if the wing were essentially flat.

A truly flat wing would require a great deal of power to sustain flight, and control would be difficult. In the early days of flight, it was discovered that an *airfoil* shape, like an elongated teardrop, would provide much more efficient lifting capability. Since then, engineers have designed many airfoil shapes to meet specific considerations.

Wings with airfoil cross sections moving through the air induce air turbulence that results in a phenomenon called "lift". This is simply a decrease in air pressure, or a small vacuum, on the top surface of the wing.

Forward motion of the wing in the air, called "air speed", is essential to provide lift. In the case of powered planes, forward motion is provided by an engine driving a propeller. This enables a plane to lift off from the ground without sharply increasing its attitude and to climb to higher altitude.

When air speed is reduced, there is a point at which the lift of the wing is insufficient to support the weight of the plane and it "stalls" or falls out of control.

At this point, you don't need to be concerned with the details of aerodynamic design. The designer of your trainer has taken care of this. -Trust him.

Air Currents and Wind

Unpowered aircraft, called 'gliders' or 'sailplanes' obtain their forward motion initially from external devices, an arm or tow rope. After they become airborne, their motion results from gravity, the tendency to fall.

A glider's upward motion can only be provided by vertical currents of air. These currents can be either 'thermals' caused by warm air rising through cooler air, or by winds deflected upward by sloping ground. In either case, the aircraft is moving forward in the air mass. If the air rises at a more rapid rate than the rate at which gravity pulls the plane downward, the plane climbs, relative to the ground.

While glider speeds are slower and their upward movement is dictated only by movement of the air, powered planes are also affected by air movement, or WIND. When the plane is flying in the same direction as the wind, it will move over the ground at a speed commensurate with the power being generated by the engine, PLUS the speed at which the wind is moving. It appears to be moving much faster. Vertical columns of air also affect R/C Planes.

When the plane flies into the wind, it is travelling at the same speed with respect to the air in which it is flying, but the air itself is moving the plane in the opposite direction. Its movement over the ground is at its normal rate, LESS the speed of the wind in which it is flying.

When the plane flies at a diagonal to the direction of the wind, it may be headed in one direction, but is being moved sideways by the wind. Unless compensation is made, the objective will be missed by a considerable distance.

Some trainer aircraft have so much lift, as related to their weight, that they will fly at very low speed. This is a significant safety factor for a novice pilot. Such planes, however, may be difficult to land when the wind is faster than their minimum flying speed.

You will learn to be sensitive to wind speed and direction. In the meantime, your instructor will help you recognize and deal with winds, both on the ground and aloft.

DIRECTIONAL CONTROL - Ailerons and Rudder

Turning on the Ground

On the ground, an airplane is like a car without brakes or a reverse gear. The throttle is used to move the plane in a forward direction and the rudder-nose wheel combination for turns. On the ground there is no up or down.

Turning in the Air

Once airborne, there is another dimension added; UP and DOWN. Turns are different, too. Like a bicycle or motorcycle, a plane must lean when it turns. In the air, this is called *banking*. If the rudder alone is used to induce a turn, the wings will tend to go into a bank, but inertia will result in more of a sideways skid.

Turns can be made more easily and with better control, using the ailerons PLUS a very little bit of UP Elevator. This is achieved by moving the stick in the direction you want to go while moving it slightly back or down. Before you attempt control in the air, practice this with your instructor on the ground where you can watch the action of ailerons and elevators closely.

The THIRD DIMENSION - Moving UP and DOWN

In the air, gravity is constantly at work to pull the plane to the ground. This introduces a whole new element to the two-dimensions we are used to in driving a car.

Climbing

The elevator is used to cause the tail to move up or down, changing the attitude of the wing. Pulling down on the stick makes the elevator go up. The flow of air against the raised elevator pushes the tail of the plane down and the nose up, resulting in a climbing attitude and the plane begins to gain altitude.

Stalls

With the throttle setting remaining constant, the engine will labor and slow down a little when the plane is in a climb. This is because the engine is literally lifting the plane. The steeper the climb, the slower the speed of the plane. If the climb is too steep, speed is so reduced that the plane loses its lift and stalls. Out of control stalls can result from poorly executed turns and other maneuvers.

In a stall, forward motion all but ceases and the plane will fall into a nose dive, heading rapidly toward the ground. Fortunately, most trainer planes will quickly reach flying speed and, if the controls are left in neutral, will tend to level out and permit the pilot to regain control, *provided the stall occurred at sufficient altitude* to accommodate that lost in the dive, with sufficient margin for recovery.

When out of control, *RELEASE THE STICKS*.

Special care must be taken during takeoff to avoid too sharp a climb or tight turns. At the moment of lift off, and for several seconds thereafter, the plane should be kept in a fairly shallow straight climb. This will enable it to increase its speed, while gaining altitude. If the climb is too steep, a stall will result and a crash is inevitable.

Descending

Moving the stick UP reverses the action and the nose goes down. The plane is in a descending attitude and loses altitude. Only very slight movement of the elevator should be tried until you are very familiar with what happens.

You will be surprised at the effects of too sharp a descent. Both engine speed and the speed of the plane build very rapidly and, unless the stick is pulled back and/or the throttle is reduced, a crash is imminent. Especially when descending, use very little stick motion. You can always add more if you need to.

Engine Speed - Using the Throttle

Engine speed has a very significant effect on the plane's ability to fly. Straight and level flight can usually be maintained comfortably at about half throttle. Climbing requires more power. If the engine is not throttled back, during descent, it can become a difficult to control dive.

If speed is reduced much below half throttle, unless the plane is kept in a slight nose-down attitude and allowed to descend, its forward motion will become too slow to maintain control. It will stall and may go completely out of control.

When you start to fly, your instructor will probably have you use full power for takeoff, half throttle for learning to fly the pattern around the field and near-idle for the final landing approach. To minimize the possibility of the engine stopping when throttle changes are made, move the control stick slowly, never abruptly.

Most trainer airplanes are what flyers call 'very forgiving'. If the plane has enough altitude, is high enough, when it stalls, simply allowing the springs to return the controls to neutral will usually result in recovery and straight and level flight. Your instructor may demonstrate this to you during your in-the-air training.

A PRELIMINARY PRE-FLIGHT INTRODUCTION to RADIO CONTROLLED MODEL FLYING

FULL SCALE Compared to MODEL Flying

There are many differences between piloting a real airplane and a radio controlled model. In a full scale plane, the pilot is *inside* and moves with the aircraft, while with radio control the pilot is outside and may be a great distance away from the plane.

It is sometimes more difficult for an experienced pilot of full-scale airplanes to adapt to the unique and very different conditions of R/C flying than it is for one with no experience.

Seat of the Pants Flying

A full scale pilot has the advantage of all of his senses to tell him what the aircraft is doing every instant of the flight. This is commonly called *seat of the pants* flying. In addition to seeing what is happening, he *feels it*. Any unusual motion is immediately sensed and can trigger an instant, instinctive reaction by the pilot to correct it.

If you have ever flown, even as a passenger in a full-scale aircraft, you know that you sense every little bump, every change in attitude, even when you can't see out of the plane.

Flying by Visual Observation Alone

The pilot of a radio controlled model must depend entirely on his vision to know what the plane is doing. As the attitude of the plane, its altitude, distance and direction from the pilot change, its visibility changes. And the farther away it is, the smaller it becomes. It is essential that the pilot *never lose sight of the plane*. Its attitude, that is whether it is flying straight and level or is climbing, descending, turning left or right must be constantly determined by the pilot. Only his eyes can tell him.

Plane Size and Color

The size and color of the plane are important factors in its visibility. The larger the aircraft, the more visible it is. Colors that blend with the sky can make a plane in the air almost invisible from the ground. Colors that contrast with the sky greatly enhance visibility of the plane. White, blue and most light colors are hard to see against common sky conditions. Bright oranges, reds and yellows contrast with sky colors and are much more visible.

Orientation

The most difficult part of radio control flying for most people is developing a sense of direction in common with the plane. You are not inside the plane, so steering becomes more difficult. When the plane is moving away from you, it is obvious which is its left or right side. You control it as if you were in it.
When your plane is coming *toward* you, however, the plane's left is your right and its right is your left. This means that to steer it you must learn to move controls as they relate to the plane's orientation, not your own.

Altitude

One of the most obvious differences between full scale and radio controlled aircraft is the determination of the plane's altitude, the distance above the ground. In a full scale aircraft, this is accurately determined by instruments that show not only the current altitude, but whether the plane is climbing or descending, as well as the rate of climb or descent. R/C pilots can only relate to these factors by visual observation.

The pilot must observe the path of the aircraft and then determine whether it is flying straight and level, or is turning, climbing or descending. His ears will help a bit, too. The sound of the engine is quite different when the plane is in a steep climb or a dive, as compared to normal flight. When the engine stops in flight, only your ears can tell you.

When you begin learning to fly, your instructor will have you flying at fairly high altitude for most of your maneuvers. This is so that, should any difficulty occur, there is more time to recover. There is an old saying among full-scale pilots that applies equally to R/C:

ALTITUDE IS A PILOT'S BEST FRIEND.

Attitude

Not to be confused with altitude, a plane's *attitude* is its position relative to true vertical and horizontal. In straight and level flight, the wings will be horizontal, that is one tip will be level with the other. The fuselage will be level from front to back and the horizontal stabilizer tips will be level with each other. In a turn, one wing tip is higher than the other. When climbing, the nose is higher than the tail. All of these are changes in attitude.

In a full scale plane, the pilot instinctively relates to his observation of the horizon to determine the attitude of the plane and learns to sense changes in attitude through his sense of balance. In addition, instruments constantly tell him precisely what the current attitude of the plane is.

GETTING READY TO FLY

You have now completed the initial Pre-Flight Training portion of your Radio Control Piloting course. Before starting at the field, take a little time to go back over this Guide to make sure that you not only understand, but are comfortably familiar with the material covered.

Discuss any questions you may have with your instructor. The hands on process of learning to fly will be much easier and faster if you have all of this material well in hand.

In the next phase, you will be working mostly at the Air Field, learning to prepare your plane for flight, fueling, starting and breaking in your engine and learning to maneuver your plane on the ground.

Be patient and thorough. You will find that a good grounding in the basics will pay big dividends as you move into actual flying.

PART II

PREFLIGHTING YOUR PLANE

GETTING READY TO TAKE TO THE AIR

Introduction

Until a new flyer has been certified as capable of performing all basic aspects of Radio Controlled Model Aircraft flying on his own, he or she should never operate a plane unless an AMA club designated instructor is on hand. When Preflighting involves operation of the engine, it should be done under the supervision of a qualified instructor or other experienced pilot using this Manual.

The FLYING FIELD

Flight training should always occur in an uninhabited area to minimize the risk of accident resulting in injury of damage to the property of others. The field runway must be flat, smooth and reasonably level and should have no nearby obstructions, such as trees and power lines. Rarely, an abandoned facility with asphalt or concrete surfaces is available as a runway for R/C flying. Usually, the field is grass covered. The grass must be kept cut very short, to accommodate the small wheels of model aircraft. AMA chartered clubs usually maintain such a field.

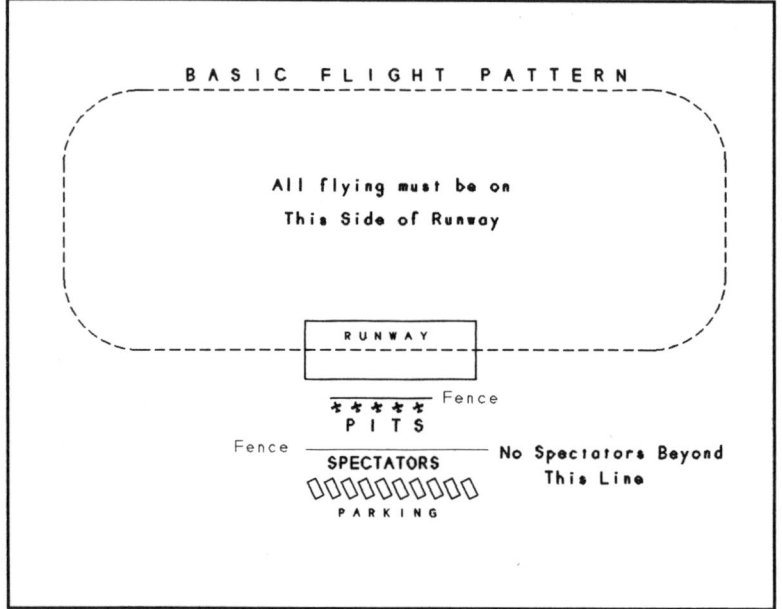

Figure 8 Typical A.M.A. Club Flying Field

Usually, several people with their planes and equipment are at a field at the same time. Fences or other barriers are frequently placed between the runway and the pit area where individual field equipment and planes are kept. This is to protect the pit area from planes possibly going out of control during takeoff or landing.

Be sure to read and become familiar with local Flying Field Rules, as well as those of AMA and observe them carefully. This is for your own safety as well as that of fellow flyers and spectators. Good pilots are always aware that they are charged with controlling a speedy and powerful aircraft in a safe manner.

If you have any questions about procedures at the field, be sure to ask your instructor or a fellow club member.

FIELD EQUIPMENT

At the flying field, you will need certain basic equipment to service, start, adjust and maintain your plane. Here, we will discuss only those things which are considered almost essential. You will probably want to add other tools and accessories as time goes on.

Field Box

While not absolutely necessary, a Field Box provides a very efficient and convenient means of storing and transporting fuel, battery and other field equipment. Usually made of plywood, with a carrying handle, it has places for a fuel container, 12 volt motorcycle battery, starter, radio transmitter, tools and accessory equipment.

The Field Box usually includes a *Control Panel* which centralizes electrical sources. Power is supplied by the 12 Volt battery. 12 volt receptacles are provided for the Starter, which is equipped with mating banana plugs. There is a 6 volt circuit with OFF-ON and reversing switches for a fuel pump. The reversing switch makes it possible to pump fuel to or from the tank in the plane. Another 6 volt circuit is provided with a rheostat to vary output voltage from 0 to 6 volts. It has receptacles for the banana plugs on the glow plug heater cords.

> NOTE
>
> *This is a Direct Current system in which current flows from positive to negative. This is called 'polarity'. When connections are made, it is essential that positive plugs be inserted into positive receptacles. Positive elements are marked with red or (+). Negative circuits are marked with (-).*

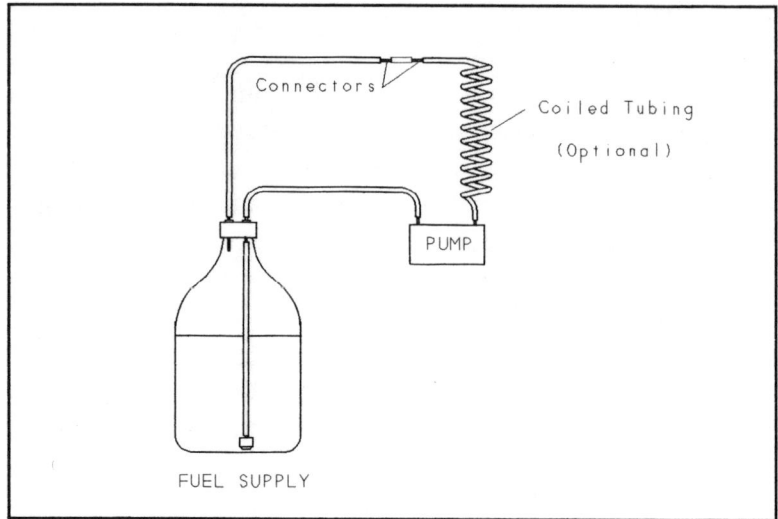

Figure 9 Field Box Fueling System

FUEL SYSTEM

A place for securing a fuel can or plastic container is also provided. An electrically powered fuel pump may be attached to the box at a convenient location and permanently wired into the control panel. Hand operated pumps may be attached to the box or to the fuel container.

Model Aircraft fuel is usually sold in gallon cans or plastic bottles. To facilitate transferring the small amount of fuel usually required to the plane's tank, a special cap with fittings that allow the connection of plastic tubes fits these containers. A short length of fuel tubing is attached to one fitting, on the *inside* of the cap. This tube, used to draw fuel from the container, should be just long enough to reach the bottom.

Another tube is attached to the outside end of this same fitting and runs to the intake side of a pump. The second fitting provides a vent and, if a tube is connected to it, a means of returning excess fuel to the container. When not transferring fuel, the other end of the tube connected to the outlet of the pump is attached to the second fitting to close the circuit and prevent evaporation.

The fuel can or bottle is commonly mounted at one end of the Field Kit.

Electric Starter

While model engines can be started by hand, this is a dangerous and difficult procedure, even when a *chicken stick* is used. An electric starter is strongly recommended and will make engine starting easier and safer.

When a Field Box is used, banana plugs on the Starter feed wires are plugged into receptacles provided on the box. If you are not using a Field Kit, spring clips on the Starter wires are attached directly to the battery. In both cases, make sure that the positive wire is connected to the positive (+) battery terminal and the negative wire to the negative terminal (-).

Model Aircraft Engine Starters are similar to automobile starters. They operate on 12 Volts and have very high torque. That means that they exert considerable force as soon as they are activated. The starter is simply an electric motor with a special flexible cup on its shaft, which is designed to fit on the spinner of the propeller, or directly on the propeller. A special switch on the side of the motor is activated by squeezing.

The cup on the starter is pressed very firmly over the spinner, then the switch squeezed. This rotates the propeller and cranks the engine, which normally starts after only a few revolutions.

The Starter is usually stored in the Field Box in a special compartment.

12-Volt Motorcycle Battery

To provide power for the Starter, a small 12-Volt wet cell motorcycle battery is required. These batteries are relatively inexpensive and are supplied with a charger. A fully charged battery will accomplish a great many ordinary engine starts. Follow the manufacturer's charging instructions carefully.

Fuel Pumps

A device of some kind is necessary to transfer fuel from its container to the plane's fuel tank. There are three common devices used for this purpose.

Electrically Powered Pumps

These pumps have a small electric motor that operates a pump. They are supplied with ON - OFF and Reversing switches. The latter are marked *IN* and *OUT* to govern the flow of fuel into or out of the plane's tank. Since they require electric power, these pumps are nearly always mounted on a Field Box and are connected at the Control Panel, where current is stepped down from 12 volts to the 6 volts required by the pump motor. Reversing the direction of the motor enables the pump to suck fuel from the plane's tank and return it to the Fuel Container. This conserves expensive fuel that might otherwise be wasted.

Hand Operated Pumps

These pumps have a handle that is rotated by hand to pump fuel. They may be mounted on a Field Box, or attached to a Fuel Container with tape.

Bulb Pumps

A rubber bulb with ball check valves at each end, these pumps suck fuel from the Fuel Container into the bulb, then force it out to the plane's tank. No mounting is required.

Glow Plug Heaters

Electricity for engine starting is supplied to the Glow Plug by a device called a *heater*. It consists of a socket that fits over the glow plug and has a battery or other power source.

Field Box Glow Plug Heater

This is a device with long wire leads with banana plugs that plug into special receptacles on the Field Box control panel. It is important that the positive lead (+) be put into the positive receptacle and the negative (-) lead into the negative receptacle.

There is usually a voltage meter and control that can be used to vary the voltage going to the heater. Follow manufacturer's instructions.

Pocket Glow Plug Heater

This is the heater preferred by most pilots. It is self-contained and powered by a rechargeable sub-C battery.

While its charge lasts a comparatively short time, it is much more convenient than the Field Box type and eliminates the possibility of lead wires becoming entangled in a rotating propeller.

Basic Tools

At a minimum, you will need a good small screw driver and a small pair of needle nose pliers and Allen wrenches to fit all Allen head screws on your engine and mounts. Also you should have hex wrenches to fit any hex nuts on your engine and the propeller retaining nut.

You will probably want to add other tools as time goes on.

Battery Chargers and Charging

All 12-Volt batteries, R/C Radio batteries and Glow Plug Heater batteries are charged in a few hours, or overnight, by chargers that plug into home outlets. It is important to follow the manufacturer's instructions carefully. It is important to remember that rechargeable batteries should not be put on the full recharge of the chargers supplied with them, until they have been discharged to a certain level.

Trickle Charging

Trickle charging allows current to flow very slowly to the battery, hence the name *trickle*. Fully charged batteries put on trickle charge will retain a full charge indefinitely. If a battery is partially discharged, a trickle charge will slowly bring it up to full charge, without damage, as would occur with rapid charging. When batteries are discharged after a full day's flying, then charged in the normal manner, they can be switched to trickle indefinitely, with safety.

Trickle Charge Adapter

A trickle charge adapter is now available for use with the chargers normally supplied with R/C radios. It is spliced into the lines between the charger and the plugs and makes it possible to change from regular to trickle charge of both battery packs by moving a switch.

Voltmeters

Knowing the condition of your radio batteries, both in the receiver and in the plane is most important. If the voltage of either one drops too low while your plane is in the air, you will lose control and a crash is inevitable. Expanded scale volt meters indicate battery condition and provide a guide as to when charging should be accomplished.

Most R/C Radio Transmitters being sold today have a built-in meter that readily tells you battery condition and indicates when recharging should be accomplished.

Your plane has no built-in meter to show the condition of the Radio Receiver battery. While you are learning to fly, one of the experienced pilots will probably have a meter he will loan you, but eventually, you will want one of your own. They are relatively inexpensive and provide great peace of mind.

DEAD BATTERIES mean DEAD AIRPLANES

Tachometers

A tachometer is a device for determining the number of revolutions per minute made by a rotating object. Ones made for use with model aircraft are helpful in refining carburetor settings to obtain maximum speed and efficiency. Usually they are of value only to more advanced flyers. Your instructor may use his to help tune your engine.

RADIO PROCEDURES at the FLYING FIELD

Every plane at the field has a radio receiver that reacts to signals from a radio transmitter to operate its controls. If two pilots operate two transmitters on the same frequency, to which the receivers of both of their planes respond, disaster is the result! Each pilot thinks he is controlling only his plane, but the other plane reacts as well, regardless of what it is doing at the moment.

R/C Aircraft Radio Channels - Frequencies

The frequencies on which any radio transmitter is permitted by law to operate are assigned by the Federal Communications Commission (FCC). A specific range of frequencies has been allocated to transmitters for the control of models of all kinds.

The allocated frequency band for all *model aircraft* is 72 megahertz. This band is further divided into sub-frequencies, or channels, assigned to transmitters for controlling individual aircraft. These channels begin at 11 and run through 60.

Most AMA Chartered clubs maintain a current list of the transmitters used by their members. It is a good idea to check with the club where you will train, before selecting the channel number of your transmitter. This will help you select a channel used by the fewest numbers of members, thus minimizing competition for channels at the flying field.

Visual Channel Identification

Most transmitter manufacturers supply a red ribbon imprinted with the words *72 Mhz Aircraft Use Only* which should be attached to the top of the antenna.

The number of the channel on which the transmitter operates must also be displayed in large letters, black on white or white on black, on the antenna.

Controlling use of Transmitters at the Flying Field

All AMA Club flying fields have a system designed to prevent two transmitters on the same channel from operating at the same time, *either in the air or on the ground.* One method uses wooden clothes pins, each with the number of a channel on it, all kept on a board at the field.

Before turning a transmitter ON, a pilot must determine that a pin with his number is on the board, indicating the channel is not being used. By removing the pin with his channel number on it, everyone else is notified that this channel is in use and no other transmitter on that channel may be turned on. The pin is attached to the base of the antenna of the transmitter using the channel, visible to other pilots.

Impounding Transmitters at the Field

As a precaution against accidentally activating a transmitter, many clubs require that every radio transmitter brought on the field be placed in a designated location and remain there until such time as its channel is free, as evidenced by presentation of the numbered clothespin or other identification symbol used at that field.

NEVER TURN ON A TRANSMITTER UNTIL YOU HAVE CLEARANCE TO USE THAT CHANNEL

A PLACE on the FLIGHT LINE

After depositing your radio transmitter, you should select a location on the flight line where you can assemble your plane and put your flight box.

PREPARING YOUR PLANE for FLIGHT

Attaching the Wing

The wing of most planes and nearly all Trainer Planes is removed for transporting and must be attached to the fuselage at the Flying Field.

Connecting the Aileron Servo

Attached to the aileron servo mounted in or on the wing at its center, is a short length of wire with a connector plug. The wing must be held close enough to the fuselage so that this plug may be inserted in the receptacle on the aileron control wire coming from the radio receiver. Make sure the connection is tight.

Care must be exercised to make sure that, when the wing is in place, this wire cannot become entangled or interfere in any way with the movement of the servo or other elements of the control system.

When removing the wing and disconnecting the servo, the two parts of the plug should be gripped and pulled.

SECURING the WING to the FUSELAGE

With Rubber Bands

If your wing is attached with rubber bands, make sure it is positioned properly, then stretch a band from a dowel diagonally across to the dowel on the opposite side of the fuselage and the opposite edge of the wing. Repeat for the other side, so that the bands are criss-crossed.

Recheck the location of the wing to make sure it is centered and perpendicular to the fuselage, then stretch rubber bands from front to back on each side, alternating sides. Use at least the number of bands recommended by the manufacturer of the kit or plane you are flying.

With Nylon Bolts

The wing of some planes is secured by means of a dowel at the leading edge being inserted into a hole in a bulkhead of the fuselage, then fastened at the rear by one or two Nylon bolts passed through holes in the trailing edge of the wing and inserted into threaded holes in a hardwood block in the fuselage. The screws should be tightened firmly.

FUELING YOUR PLANE

Disconnect fuel lines from the muffler and carburetor. Attach the tube from the discharge side of the pump to the tube you disconnected from the carburetor, using a short length of brass tubing as a connector. This will direct fuel to the on-board tank. The tube you disconnected from the muffler will act as an overflow when the tank is full. If you have a tube to return fuel to your main fuel container, join it to the muffler tube with a connector.

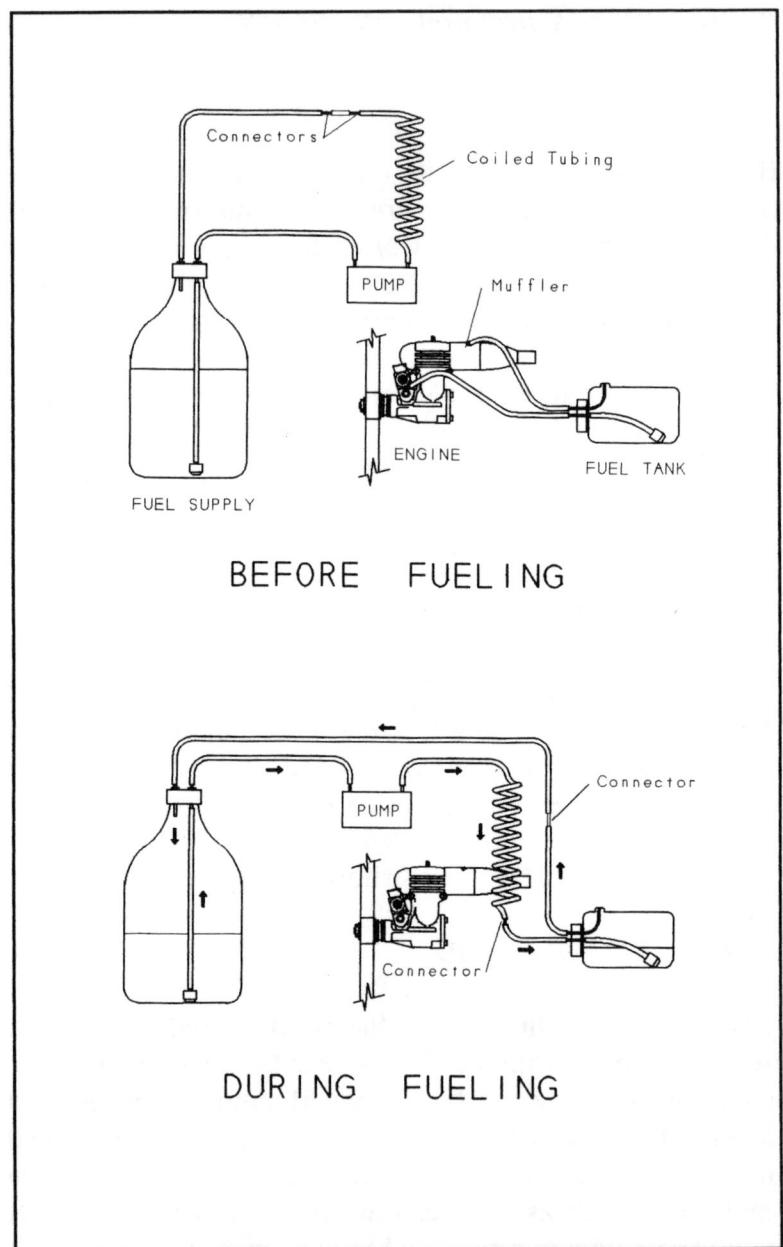

Figure 10 Fueling System

If you do not have a return tube, the muffler tube should be held over the side of the engine compartment, so that overflow fuel is directed away from the plane. It is strongly recommended that you have a return line connected to the fuel supply bottle.

Set the fuel pump direction switch to OUT and the main switch to ON. It may take a few seconds for the pump to prime itself and start moving fuel to the tank. The fuel tubes are translucent and the fuel is colored, so you will be able to observe the fuel as it starts to flow.

The plastic fuel tank in some planes is visible during filling, so that the fuel level can be watched and pumping stopped before the tank overflows. If this is not the case with your plane, you will have to watch the overflow tube and stop the pump as soon as you see fuel coming out.

After fueling is complete, reattach the tubes from the fuel tank to the carburetor and the muffler. Join the two brass connector tubes in the line between the pump and the supply tank with a short length of silicone fuel tubing. This closes the line and eliminates evaporation of fuel.

STRUCTURAL CHECK

Inspect the wing, fuselage, all elements of the tail assembly, the landing gear, engine, propeller and spinner, as well as the attachment of clevises to control horns of elevator, rudder and throttle, to make certain nothing is loose. Make any corrections necessary.

Check engine mounting bolts and landing gear screws. Tighten them as necessary. Make sure vertical and horizontal stabilizers are firmly attached.

ACTIVATING the RADIOS

Verify that your radio channel is not in use. Pick up your pin, or other identifying token and place it on your transmitter antenna. At some fields, you may be required to leave your AMA membership card in place of the token. This identifies the current user of each channel. Once you have completed this procedure, that channel is yours and you may proceed.

CHECK RECEIVER BATTERY CONDITION

It is essential that the battery in your plane have sufficient power to operate for at least the duration of one flight. The only way to be sure of this is to check it with a special expanded scale volt meter. Your instructor will probably have a meter and will be glad to check for you.

TURN ON the RECEIVER in the PLANE

If your receiver battery checks out OK, move the radio switch on your plane to ON. You will probably hear a small whirring sound and, if you are watching the control surfaces, you may see them move. This is the radio centering the servos, which may have been moved by pressure on the control surfaces during handling and transportation.

TURN ON the TRANSMITTER

When the transmitter is turned ON, there may be some more movement of the servos in the plane, as they move to match the settings of the transmitter controls. Most transmitters have a meter that operates when the switch is ON. The meter may show signal output power, but in

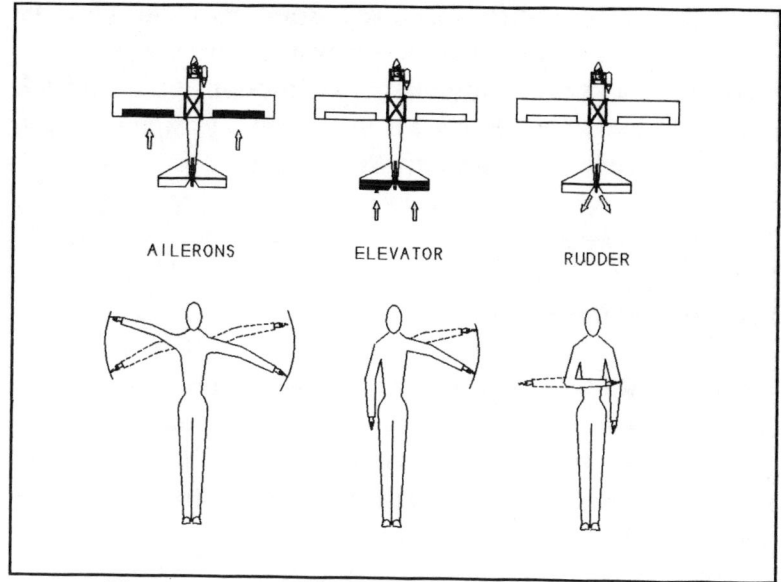

Figure 11 Radio Range Check Hand Signals

more recent models it is an Expanded Scale Voltmeter that indicates battery condition.

In either case, when the indicator hand falls into the red zone on the dial, there is insufficient charge for safe flying and the batteries must be recharged before flight.

NEVER ATTEMPT A FLIGHT UNLESS BOTH TRANSMITTER AND RECEIVER BATTERIES ARE CHARGED

RANGE CHECK THE TRANSMITTER

It is most important that you make sure the receiver in your plane is responding properly to signals sent by your transmitter.

With both receiver and transmitter switches ON, carry the

transmitter at least fifty feet away from the plane, with the *antenna DOWN*. Have someone watch your plane as you move the sticks in all directions, one motion at a time, and, using the hand signals shown in Figure 11, signal to you when the control surfaces are responding.

STARTING the ENGINE

Breaking In the Engine

Nearly all Radio Control Aircraft Engines require breaking in under conditions specified by the manufacturer. Usually this involves running the engine with a rich mixture for short periods of time, alternating with lean mixtures for short periods.

The throttle is run partially open for most of this procedure, but occasional short runs at high throttle may be recommended. This process is repeated until two or three tanks of fuel have been burned.

Many experienced pilots, with certain makes of engines, after adjusting the engine prefer to run the engine slow, then fast, then slow, etc., until about a tankful of fuel has been burned. The balance of breakin is accomplished in the air, using a comparatively rich mixture for several tanks full of fuel.

It is strongly recommended that you have an instructor assist you with this important task. His experience and knowledge of engines can help avoid damage to the engine.

Some pilots find that certain makes of engine do not require breakin and can be flown without it. While it may take a little longer, a breaking period is normally recommended.

STARTING the ENGINE

Turn the Radios ON and Set the Throttle

Both the transmitter and the receiver must be on before starting the engine. Use the stick on the transmitter to set the throttle at or slightly below the midpoint.

Keep the radio transmitter at hand during starting, so quick throttle adjustments can be made.

Holding the Plane Against Moving

Two forces act on the plane, when starting the engine. The fuselage must be held firmly to resist the *backward* force necessary to create friction between the starter and the propeller spinner. As soon as the engine starts and the starter is withdrawn, the propeller begins to pull the plane ahead. The fuselage must then be firmly restrained against *forward* movement. Holding the fuselage tightly, just ahead of the wing, is the most commonly used means of resisting both forces.

Priming the Engine

Be sure the GLOW PLUG HEATER is *NOT ATTACHED*

Priming simply means drawing some fuel into the cylinder before starting the engine. This is accomplished by holding a thumb over the carburetor intake opening and rotating the propeller *in a counterclockwise direction* several times. This sucks raw fuel into the cylinder.

Occasionally, especially in cold weather, introducing a few drops of fuel into the carburetor throat may be necessary.

Connecting the Glow Plug Heater

If you are using a heater deriving power from the Field Box battery, make sure the leads are properly connected and the voltage set correctly at between 1½ and 2 volts. If using a battery powered heater, make sure the battery is charged. If the engine does not start, replace the plug.

The glow plug heater is attached to the glow plug just before you are ready to start the engine. Make sure it is securely snapped in place, to assure good contact.

Using the Starter

Plug the banana plug leads of the starter into the sockets on the control panel of the Field Box. Make sure the *positive* and *negative* plugs are in the proper receptacles. (*Red to red, black to black.*)

CAUTION: Make certain that the starter lead wires are positioned well away from the propeller when the starter is in use and as it is withdrawn and set aside after the engine starts. If the wires become entangled with the propeller, injury or serious damage is almost inevitable.

The starter switch is located under a rubber bar which should be on the bottom of the starter as you hold it in your hand, with your fingers loosely over the bar. Press the rubber socket on the shaft of the starter *very firmly* over the center of the propeller spinner, *then* squeeze the switch bar.

Be prepared for a sharp jerk on your wrist as you resist the torque of the starter when it spins the propeller. As soon as the engine fires, quickly but carefully withdraw the starter from the spinner and release the starter switch.

BEWARE of the SPINNING PROPELLER !

Post-Starting Carburetor Check

To make as sure as possible that the engine will perform safely in the air, it is a good idea to make a few tests before takeoff. To begin with, your instructor will assist you in all these activities.

Speed Variation

First, check the idle setting. With the throttle control fully down, the engine should run smoothly at slow speed. Adjust as necessary, restarting the engine if it stalls.

After the idle setting performs properly, move the throttle to full open, then back to idle a couple of times. This should be a fairly rapid, smooth motion. The throttle should never be moved sharply. Make any carburetor mixture adjustments necessary to produce a slightly rich mixture producing a little smoke at full throttle.

When the transition from fast to slow and back can be accomplished smoothly and without the engine coughing or stopping, the carburetor should be properly tuned.

Attitude Test

With the throttle set at full throttle, hold your plane in the air so that the fuselage is almost vertical. If the engine slows down or stops, there may be fuel line problems, or further carburetor adjustment may be necessary. Repeat this procedure as often as necessary to eliminate any problems. When the engine runs smoothly in this attitude, reduce the speed to idle and taxi to the flight line.

Your Plane's MAIDEN FLIGHT

AMA Safety Rules require that the first flight of any new aircraft, or of one that has not been flown for some time, or has undergone major repairs, be made by an experienced pilot. Thus, the first flight of your plane will be entirely under the control of your instructor. During this flight, he will demonstrate and explain a number of important basic aspects of R/C flying.

Stay close to the instructor during the entire flight and pay close attention to his or her comments.

Rules and Courtesy While on the Field

Pilots taking off are normally permitted to stand on the runway, directly behind their plane, during takeoff. As soon as the plane is airborne, the pilot should move off the runway to the area along its edge designated for pilots during flight. A student is permitted to accompany a pilot demonstrating flight techniues and should stay close to him.

Your instructor will demonstrate proper procedures for takeoff, while your plane is in the air, deadstick (no power) landings, recovery from faulty maneuvers. A well-trimmed plane should recover from most out of control situations, if the controls are returned to center.

Trim All Controls

During your plane's first flight, your instructor will experiment with the plane's reaction to his instructions, issued by movement of the sticks. With all controls in neutral, he will observe the plane's attitude and will adjust the trim controls to bring it into straight and level flight.

Check Plane's Performance

He will experiment with more radical maneuvers to check the plane's response and its performance at various engine speeds and in various strenuous maneuvers.

This is a safety precaution to verify the plane's airworthiness and uncover any flaws before you take control. Your instructor may try several landings and takeoffs. He should explain what he is doing as he goes along.

Demonstrate the Local Field Pattern for Takeoff, Normal Flight and Landing

It is important that you become familiar with the safety rules, patterns and procedures relating to all flying activity at the flying field at which your training takes place. Think of the other fellow at all times and be courteous, not interfering with his activities.

CONTROLLING YOUR PLANE ON THE GROUND

Learning to Taxi

Steering a plane on the ground is called taxiing. It is accomplished by moving the left stick on your transmitter right or left in the direction of the turn, while at the same time moving the same stick up or down to control engine speed.

Most R/C taxiing is done on grass fields where the grass, even when cut short, is high enough to provide a real obstruction to the small wheels of your plane. This is especially noticeable when starting from a dead standstill.

Tufts of grass and small depressions act like wheel chocks and it takes a fair bit of power to pull the plane over or out of them. When the plane starts moving it builds inertia, so the mass of the plane in motion helps the engine move over the grass.

Generally, taxiing winds up being done with short bursts of power, as needed. Care must be exercised so that you don't reach speeds sufficient to cause the plane to take off when you don't want it to.

Practicing Taxiing

All taxi practice should be done under the supervision of your instructor, or another experience pilot.

Clearing Your Radio Channel

Remember that even though you have no intention of flying, your radio must be used to control the throttle and the rudder. You must go through the same procedures of making sure the channel on which your radio operates is free and reserving it for yourself by placing the pin on your antenna, or whatever other procedure your club has established.

Failure to do so can result in your radio signals being received by a plane in the air and loss of control by its pilot.

Avoid Interfering with Other Pilot's Activities

Many clubs reserve certain times at their fields for use exclusively by those learning to fly. Taxiing, like other aspects of learning to fly can best be accomplished at these times, but don't hesitate to practice at other times, as long as you don't interfere with the activities of other pilots.

Planes landing and taking off have priority. You should be careful, when practicing taxiing to be aware of the activities of other pilots and to practice in an area of the field away from the runway, unless there is no one else at the field who wants to fly.

Using the Throttle

With the engine idling, spot your plane on the ground where it is in the clear and away from any obstructions.

Experiment with the throttle to see how much power it takes to move your plane. Be prepared to throttle back instantly. The plane will have a tendency to jump away very rapidly. Apply short bursts of power, gradually lengthening them until you can keep the plane moving continuously for some distance.

Steering - Taxiing

Practicing taxiing on the ground will help accomplish an important aspect of learning to fly R/C, *reversing your thinking when the plane is coming toward you.* You should become proficient at taxiing before you take to the air. It will also accustom you to controlling two things at the same time, *turning* and *speed control*.

You will find that the engine must be speeded up to pull the plane over hummocks of grass and other irregularities in the field surface, then quickly slowed down as the plane begins to gain momentum. This demands close attention and up and down movement of the stick to control the

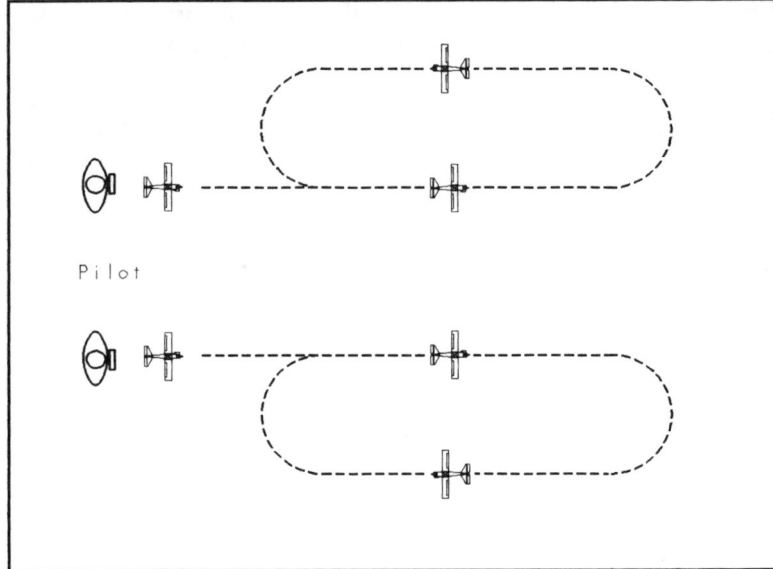

Figure 12 Racetrack Practice Taxi Pattern

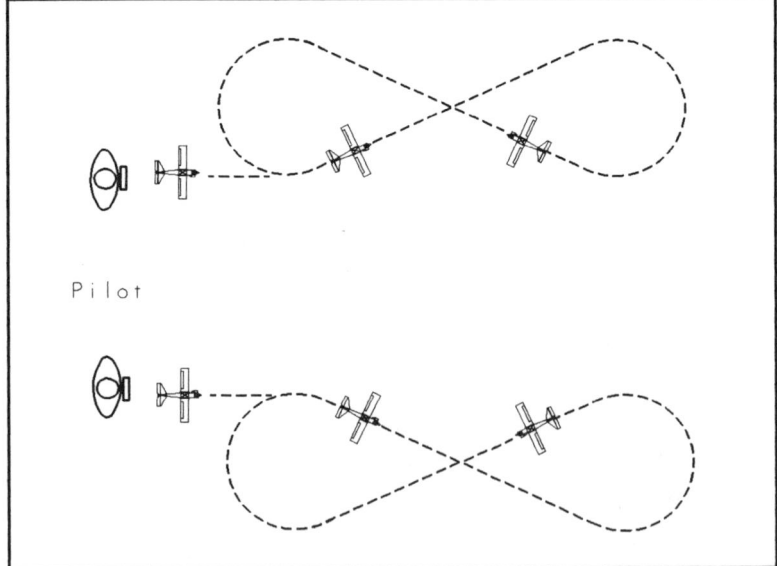

Figure 13 Figure 8 Practice Taxi Pattern

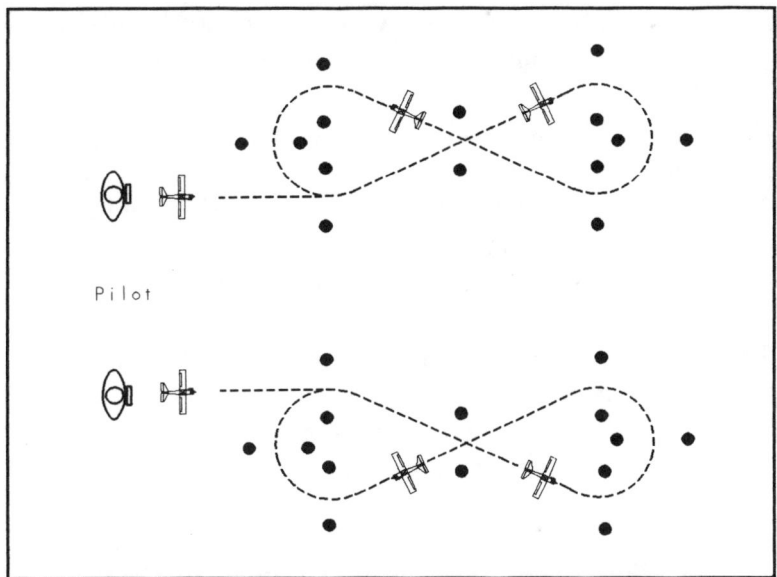

Figure 14 Figure 8 Advanced Taxi Practice Pattern

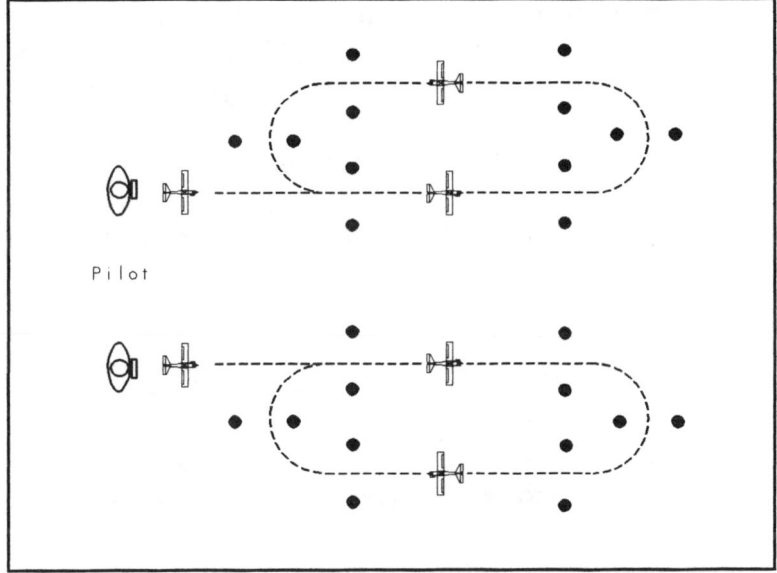

Figure 15 Racetrack Advanced Practice Taxi Pattern

throttle. At the same time, you must control the plane's direction by left and right movement of the same stick.

Practice turning by moving the stick both left and right to get a feel of the amount of stick movement required. Turning can, of course, only be accomplished while the plane is moving. Try making small changes in direction, then full left and right turns and complete 180º turns.

Take the plane some distance away from you, then turn it around so it is coming toward you, practice turning left and right as it moves in your direction. Do this repeatedly, until you instinctively reverse your control actions when the plane is coming your way.

As you become more comfortable with ground maneuvering, try following a predetermined course, as shown in Figure 12 and Figure 13. As you become more proficient, try the same patterns, but with gates formed by plastic gallon jugs or two liter soft drink bottles. They should be spaced about two feet farther apart than your wingspan.

SHUTTING DOWN and PACKING UP

When you have finished the day's activities at the field, certain chores must be accomplished before you can go home.

MAKE SURE ALL SWITCHES ARE OFF

Any fuel remaining in the tank of your plane should be pumped back into your main fuel tank, or otherwise removed. The Wing must be removed by taking off the rubber bands or loosening the screws.

You will find that the exhaust from your engine has left an oily coating on many of the surfaces of your plane. This must be thoroughly removed, using a spray detergent, such as 409, and paper towels. Examine all surfaces, it is surprising where this oil can find its way, and clean any found contaminated. Dispose of used paper towels and other trash in containers provided at the field, or take them home with you for disposition.

Disconnect your starter and glow plug heater from the field box, coil the cords and store them in the field box. Check the area you have occupied on the flight line for tools, parts or waste materials. Leave it clean and free of any kind of debris.

WATCHING OTHER PEOPLE FLY

You can learn a great deal by watching other people as they start, taxi, take off, maneuver in the air and land their planes. It is well to bear in mind, however, that, as in other activities, there are good and not so good pilots. Don't try to copy those who are erratic or abrupt with their controls.

You will soon be aware of the good pilots, those whose movements are smooth and well coordinated, whose takeoffs and initial climb follow the established field pattern. They may be outstanding aerobats when in the air, but the real test of a good pilot are the consistency and quality of landings and takeoffs.

Before long, you will become aware of mistakes other people make. Make a mental note of what was done wrong and determine to avoid similar practices.

PART III

PRIMARY FLIGHT TRAINING

PUTTING YOU IN CONTROL IN THE AIR

Introduction

For the first time, you are about to actually take over control of your airplane while it is in the air. Remember that your instructor is always right there to take over if you should happen to make a boo boo. Relax, take it easy and really enjoy this unique experience !

YOUR FIRST EXPERIENCE in CONTROL

Before this flight, your instructor or another experienced pilot should have tuned your engine, helped you break it in, put your plane in the air and trimmed it. Before you take control in the air, it is essential that your plane be in the best possible flying condition.

Takeoffs and Landings

These maneuvers, the most critical in flying, should be attempted by the beginner only after he has learned to effectively manipulate the controls and the plane, in the air. Until that time, the instructor will take the plane off and land it.

YOU and YOUR INSTRUCTOR

While some people have learned to fly R/C without an instructor, this experience has nearly always involved many crashes, the frustration of rebuilding or replacing planes and perhaps radio elements and substantial cost. The experienced pilots who act as instructors in AMA Clubs do so as a service to the Club and to the student in an effort to minimize the cost of learning and to help you learn faster.

Your instructor is dedicated to the task of helping you develop the skills and confidence necessary to enable you to fly Radio Controlled Planes with assurance and safety. He is concerned about your safety and that of your plane. He cannot perform miracles, however. In the unlikely event of an accident during your training, your instructor will probably feel worse than you do. Don't blame him.

Especially during this early flight phase, don't hold back. Ask your instructor any questions that bother you and be sure to let him know how you feel, as you progress. When you feel ready to take the next step, tell him so. If he agrees that you are ready, he will encourage you to try. If he thinks you need more practice, he will tell you. In most cases, you will both know when the time is right.

Communicating With Your Instructor

Things happen very quickly in the air, you can't just pull off to the side of the road and stop, and it is essential that someone be in constant control of your plane. Make sure you and your instructor agree on exactly what is planned for the flight. Then keep in verbal contact throughout the flight.

It is strongly recommended that the terminology be kept as uncomplicated as possible. Most students are unfamiliar, or only newly acquainted with many of the terms that will later become second nature. In some cases new terms are the only alternative. It is a good idea to rehearse instructions that will be given, just before the flight. Here are a few examples:

Left Aileron - Right Aileron

> Move the right stick to Left or Right. This instruction is used when the instructor wants you to operate *ONLY* this control.

Left Turn - Right Turn

> When *on the ground*, this means to move the left stick right or left to steer the nose wheel.

> When *in the air*, it means making a coordinated turn, using ailerons and elevator and, sometimes rudder controls.

Elevator FORWARD - BACK

> Here there is danger of misunderstanding. Moving the Right Stick DOWN raises the elevator, causing the tail to go down and the nose to go up, causing the plane to climb or go UP. Moving the stick UP lowers the elevator, raising the tail and pointing the nose DOWN.

> It is recommended that the terms used in Full Scale flight be used here, substituting *ELEVATOR (Stick) BACK* for UP and *ELEVATOR (Stick) FORWARD* for DOWN.

When landing a plane, just before the wheels touch down, it is customary to pull back on the elevator stick, bringing the NOSE UP and causing the plane to settle to the ground. This is called a *FLARE* maneuver. Instructors are urged to say "*ELEVATOR BACK*", rather than "Flare", so the student does not have to translate.

Throttle - UP, HALF, DOWN

There is seldom any confusion about throttle instructions. "Slow, or Idle, Full Throttle, Half Throttle", etc. are seldom misunderstood.

The IMPORTANCE of ALTITUDE and DISTANCE

ALTITUDE and SAFETY

In both full scale and model flying, altitude translates to TIME. If something goes wrong while the plane is near the ground, there is little if any time to regain control and avoid a crash.

The higher the plane, within reason, the more time there will be to recover and minimize the possibility of an accident.

Until your instructor is satisfied that you have learned to control your plane well, he will want you to maintain an minimum altitude of several hundred feet while you are learning to control your plane smoothly. Above this minimum altitude, you may climb, descend and perform other maneuvers.

DISTANCE - Keeping Your Plane in Sight

It is most important that you keep your plane near enough so that you have no difficulty seeing it at all times. This means being able to tell the attitude of the wings and the fuselage and whether it is going up or down, turning left or right.

Remember that you must depend completely on your vision to tell you what to do to control your plane. Turn it back toward the field long before it becomes hard to see.

Altitude involves distance as well. Don't let your plane go so high that you have difficulty seeing it.

YOUR Vision Governs

While your instructor will try to help keep you from straying too far away, or too high, he cannot know exactly what YOU are seeing. Vision, particularly distant vision, varies significantly between individuals. Your instructor may be able to see the plane distinctly at a much greater distance than you can. If you cannot see the plane clearly, tell the instructor. When you have the controls, keep the plane well inside the distance at which YOU can see clearly and sharply.

It is important that you discover early on how to keep your plane within the range of your vision.

TRANSFERRING CONTROL
Between Instructor and Student

In the air, your plane is always moving. It cannot stop and wait for a control command. While your instructor will qhave trimmed it so that, with hands off the control sticks, it will fly straight and level, it is moving forward at a high rate of speed. For safety, transferring control from instructor to student and from student to instructor must be accomplished smoothly and quickly.

Both must have the plane clearly in view.

Using a Single Radio Transmitter

The most usual training system involves only the student's radio transmitter. Except in emergencies, the plane should be in straight and level flight at the time of transfer.

Whenever control is transferred from instructor to student, or vice versa, the person using the transmitter will say "I'm ready to turn control over to you. Are you ready?" Only when the other person answers affirmatively, is the transmitter quickly handed from one person to the other.

Using Two Radio Transmitters and a "Buddy Cord"

More and more R/C Radio Transmitters are now equipped with a special switch and a receptacle to receive the plug on a "Buddy Cord". The student's transmitter, on the same frequency as the plane receiver, is used to actually control the plane. Its antenna must be fully extended.

A second, similarly equipped transmitter can be of any frequency, because it will send no radio signals. It is connected to the first one by a cord, commonly called a *Buddy Cord*. When the controls of the second transmitter are operated, a message is sent through the cord to the first transmitter which then sends a radio signal.

The antenna of the second transmitter should be collapsed and the switch must be kept in the OFF position at all times. In this way, it cannot accidentally send out a signal, except through the first transmitter.frequency.

Each transmitter has a spring loaded toggle switch mounted on the top of the case at the left side. Only the switch on the transmitter matching the plane's frequency is used. When in its normal position, movement of its control sticks will be transmitted to the plane's receiver. This transmitter is used by the instructor.

When the toggle switch on the instructor's transmitter is held to the left, his transmitter is disconnected and the signals from the second or student's transmitter are transmitted by the instructors transmitter. The instructor can regain control instantly, simply by releasing pressure on the switch and allowing it to return to its normal position.

Except in an emergency, the controls of both transmitters should be in the same position when control is transferred. Normally, the plane will be flying straight and level, with the elevator, rudder and aileron controls in neutral, that is with no pressure applied to them.

The throttle control is not spring loaded and stays wherever it was last set. It is suggested that, to avoid surprises at the instant of control transfer, the person who is not controlling the plane keep the throttle stick at half throttle. The person relinquishing control should then move his stick to half throttle and allow the plane to stabilize at this speed before control is transferred.

If there is some difference in the throttle setting of the transmitters, it will be evident by a change in engine speed and the sound it generates. The person assuming control can make adjustments without endangering the plane.

An important advantage of the half-throttle at transfer rule, is that the slower speed of the plane is less likely to cause a sharp maneuver if the student accidentally over controls before getting the feel of the controls.

EMERGENCY TRANSFER of CONTROL

It is almost inevitable that, at some time or another, a student will find himself in a situation where he is unable to control his plane. To avoid a possible crash, it is essential that the instructor take over control as rapidly as possible.

While the instructor will normally keep his eye on the plane and will recognize an emergency situation and take the transmitter away from the student and try to regain control. Especially when he has begun to have some confidence in the student's ability, the instructor may be distracted for a short time.

As soon as he sees he is in trouble, the student should shout "TAKE OVER!" and immediately hand the transmitter to the instructor.

When there are two transmitters and a buddy cord, releasing the switch can be accomplished almost instantly.

TAKING OVER the CONTROLS
for the FIRST TIME

Now, you are in for a wonderful new experience. After taking the plane to an altitude of several hundred feet and positioning it so it can safely fly straight and level for some distance, your instructor will transfer control to you.

Don't expect any miracles. Learning to fly Radio Controlled Models, like full scale aircraft, requires learning basic principles followed by *practice, practice, practice.*

Your Instructor's Flight Plan

Before takeoff, you and your instructor should discuss what will happen during your first flight. The plan should include learning to turn both left and right, while maintaining altitude, practicing following the designated field pattern and perhaps other maneuvers.

Getting the FEEL of the Controls

Your instructor will probably ask you to make a turn in one direction or the other. Keep a light touch on the control stick and move it only very slightly, observing the plane to see the results of your actions. Avoid jerky motions.

With the exception of the throttle control, the sticks are spring loaded to automatically return them to the center, or neutral, position. While you should maintain finger contact with each stick, it is most important that your touch be light enough to let you know when the sticks are centered and there is no pressure on you fingers from the springs.

You may want to practice with the transmitter switch OFF, moving the sticks in all directions and allowing them to return to center, maintaining constant finger contact. Doing this repeatedly will develop a finger sensitivity that will give you a good feel of the controls.

Basic Maneuvers

As you are ready, your instructor will help you learn and various maneuvers in the air are alternating right and left turns while maintaining altitude, figure eight turns, gentle climbs and descents and the same maneuvers at varying throttle speeds.

Landing patterns, including throttling down and descending should be practiced, including fly-bys, low but comfortably above the runway.

TAKEOFF

After considerable practice in the air, when your instructor feels you have gained enough familiarity with the controls, he will ask if you think you re ready for your first takeoff. If you feel you are ready, he will help guide you through this critical experience.

The most important thing on takeoff while on the ground is to maintain direction, using the nose wheel and rudder. When the plane leaves the ground you must switch to aileron control to keep the wing level. Apply slight up elevator at lift off to maintain a gentle climbing attitude.

Hold a steady course as the plane climbs, keeping the wings level, until it is at least a hundred feet off the ground, then use aileron control to turn in the direction of the field pattern in use, maintaining a climbing attitude in the turn.

Your instructor may suggest that he land and let you take off a number of times, until he feels you have mastered the basics. At that time you will be making smooth takeoffs and climbs into the pattern on a regular basis.

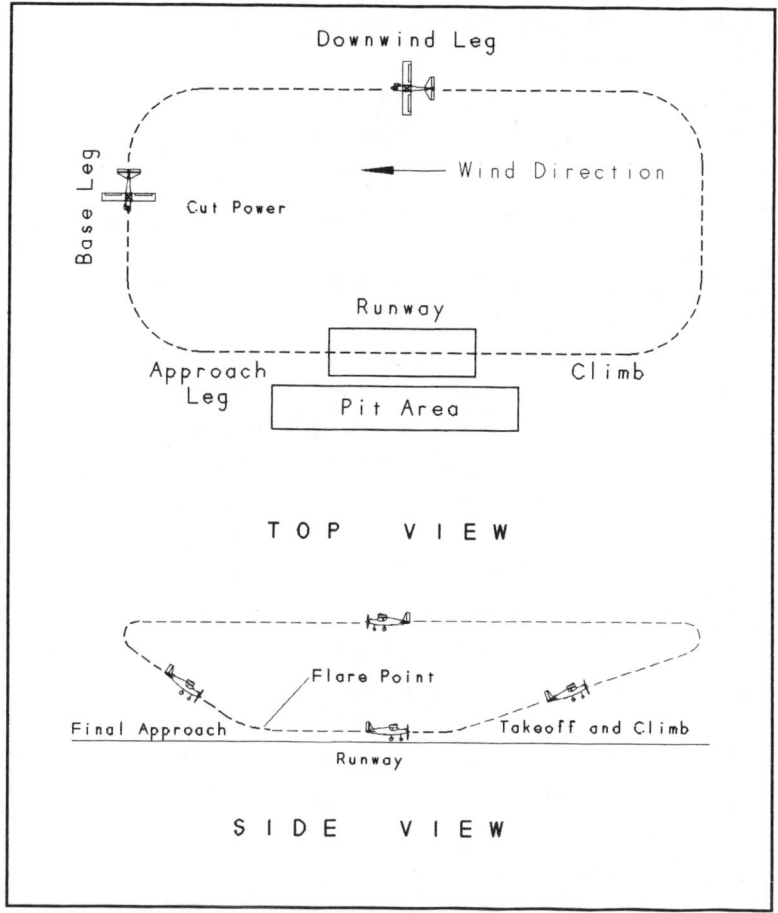

Figure 16 Practice Take Off and Landing Pattern

Carefully watching and listening to your instructor during approach and landing will help ready you for your own first landing.

LANDING

Landing is the most difficult of all maneuvers to accomplish safely. When teaching landing, your instructor will probably have you practice making landing approaches, These will include throttling down and gently descending toward the end of the runway, lining up with the runway then throttling up before actual touchdown, maintaining a straight and level course above the runway. After climbing to pattern altitude again, another approach can be made.

Throttling Back - the Gliding Approach

The point at which the throttle is cut back to idle will be in the base leg. As the pull of the engine is removed, a little down elevator may be necessary to keep the nose slightly down. In a nose-down attitude, gravity will provide sufficient force to maintain forward speed and lift. The steeper the descent, the faster the plane will move. Ideally, the point at which the throttle is retarded will enable a gentle glide, ending over the end of the runway. You will learn how to determine this point by your instructor, then by trial and error, then practice.

It is likely that many practice approaches will be suggested before an actual touchdown is made.

Flaring at Final Touchdown

When the plane approaches the end of the runway, with the engine at full idle, it remains airborne only because its nose is slightly down and gravity is pulling at it. When it is over the runway and only a foot or so above it, a slight backward movement of the elevator stick lowers the tail and raises the nose, causing the plane to lose speed and begin to stall. This is called *flaring*. The plane at this point will drop to the earth, the landing gear absorbing the shock of landing. Trainer Plane landing gear are designed to accept many very rough landings without permanent damage.

Maintain Communication with Your Instructor

Let your instructor know how you feel about what you are doing. Don't hesitate to ask questions. *The only dumb question is the one that is never asked !*

Flight Duration

While it may seem like forever, most trainer planes can remain in the air for from ten to fifteen minutes. This will vary with throttle setting. If you always fly at half throttle, your fuel will last longer than at full throttle. Small quartz kitchen timers, available at K Mart and other similar stores, can be set to sound a warning note at the end of a preset period and can be attached to your transmitter.

PRACTICE, PRACTICE, PRACTICE !

Typical A.M.A. Chartered Club

RULES for SAFE FLYING

These rules govern activities of any and all persons on the property operated by the Club as its Flying Field. They are designed to insure the maximum possible level of safety and enjoyment for both flyers and spectators. Failure to comply will result in loss of flying privileges and/or ejection from the field, at the sole discretion of the Safety Officer.

ALCOHOLIC BEVERAGES and ILLEGAL DRUGS ARE STRICTLY PROHIBITED

PERSONS UNDER THE INFLUENCE OF ALCOHOL OR ILLEGAL DRUGS WILL BE SUMMARILY EJECTED

A.M.A. MEMBERSHIP

Only persons holding a card evidencing current membership in A.M.A. (Academy of Model Aeronautics) may fly at the Club Field. At the sole discretion of the Safety Officer, a person holding such membership, but who is **not a member of this club,** may be permitted to make not more than three solo flights under the direct supervision and control of a recognized club Instructor. Before any further flying, such persons must become members of the Airmasters R/C Club.

Members of other A.M.A. affiliated clubs may be invited to fly at the Airmasters Field on special occasions, when authorized by an official action of the membership. They will be expected to fully comply with these Rules and those of A.M.A.

Members of this club may invite qualified pilot members of other clubs to fly at Airmasters Field as their guests. The member will be held responsible for the conduct of such guests.

PERSONS IN PIT AREA

Only those members, or authorized guests, who are actually involved in operation of aircraft are permitted inside the fence separating the Pit area, or Flight Line, from the Parking Lot.

DESIGNATION of TAKEOFF and LANDING DIRECTION

The Safety Coordinator shall designate, on the basis of current wind conditions, the direction in which aircraft will take off and land. He will constantly monitor changes in wind direction and will make sure all flyers are aware of the current designated direction.

Except in emergencies, the designated direction must be strictly observed by all pilots during takeoff and landing.

FLIGHT PROCEDURE

Before any flight, each pilot should check the following list and each item complied with.

To prevent accidental interference with control of an aircraft in flight by another pilot, RADIO TRANSMITTERS MUST BE OFF at all times, except when the Pilot has positively confirmed that the channel he will use is clear and has claimed it for his use in accordance with these procedures.

The plane must be inspected for structural and control reliability, including the attachment of control rod clevises to control surface horns.

Radio Channel Designation

Each Radio Transmitter must have a red ribbon Flag indicating it is for use on the 72 Mhz aircraft band. In addition, readily visible numbers indicating the channel on which it is transmitting, must be **attached to the antenna**.

Frequency Control Board (This is only one of several control devices.)

The Club has provided a closed cabinet, at the entrance to the Flight line, containing wood clothes pins, each bearing the number of an authorized R/C channel. These pins are attached to rods and are placed in front of a number painted on the back of the case, corresponding to that on the pin.

Obtain Proper Channel Designator Clothes Pin

Håving determined that there is no other transmitter operating on your frequency, remove the pin bearing the number corresponding to your transmitter frequency from the Frequency Board and Attach it to your Antenna.

Verify Correct Operation of Control Surfaces

Turn Transmitter and Receiver ON and operate Control Sticks while observing movement of Control Surfaces. Make any necessary corrections.

Range Check Transmitter

Operate Transmitter controls while moving away from your aircraft, having an observer confirm that surfaces operate properly from a distance of at least 50 yards. (Use arm signals shown on p. 51)

Starting and Running up Engines

When starting engines, the aircraft must be facing AWAY from spectators, parked cars and others in the Pit Area.

Taxi Procedure

Before taxiing out of the Pit area, the Pilot should announce his intention by calling out, "COMING OUT!" When he is sure that all persons on the field are aware of his intentions, he may proceed to taxi out to takeoff position.

Persons on Runway

Only the Pilot and one observer or student is permitted on the runway and then only during actual takeoff and landing.

Takeoff Procedure

When ready to take off, make sure other aircraft in flight are well clear, then call out "**TAKING OFF!**" When others are aware of your intentions, proceed with takeoff.

Following takeoff, immediately move to the edge of the runway and continue control from the sidelines.

In-Flight Activities

All normal flying must be conducted in the air on the opposite side of Runway from the pit area.

Landing Procedure - Normal

When preparing to land, enter downwind leg and observe to make sure no other aircraft are in the pattern. When beginning final approach, call out "**COMING IN FOR A LANDING!**" and if all is clear, proceed to land.

Deadstick Landing

In the event of engine failure, immediately call out "**DEADSTICK!**" All other aircraft must immediately make way and give the unpowered aircraft a clear path to the runway.

This procedure must also be used by gliders when on final approach.

NEW or Repaired AIRCRAFT

New aircraft which have not previously been flown, or have been repaired or modified, should be inspected by an experienced member before making their first flight.

FLIGHT TRAINING

Qualified members have been designated by the club as Volunteer Flight Instructors. They are available to train New Members desiring to learn to fly R/C Model Aircraft. <u>Instruction at the Club Field may be conducted only by these designated individuals.</u>

Trainees may fly only under the direct control of an Instructor. Until the student has demonstrated to the instructor that he/she is capable of safely flying his aircraft in a simulated landing pattern, only the instructor shall perform takeoff and landing.

FIRST SOLO FLIGHT

When a student is considered by his Instructor to be proficient enough in controlling his aircraft to perform solo takeoff and landing and he is ready for his first solo flight, this intention will be announced to all flyers present at the Flying Field.

<u>Before his aircraft is placed on the runway, all other aircraft should be on the ground, all radios OFF and no engines running.</u>

Initial solo flights shall be under the direct supervision of the Instructor and will include takeoff and as many touch-and-go landings and takeoffs by the student as he deems desirable.
The Student's qualifying solo flight will be conducted under the same conditions <u>with no other aircraft flying</u>. Upon satisfactory completion of this flight, the Instructor may certify the student as a qualified Pilot, in accordance with A.M.A. criteria.

When the these solo flights are completed, the Instructor shall so notify other flyers that the Field is again open for normal operation.

NO PERSON WHO HAS NOT BEEN CERTIFIED AS A QUALIFIED PILOT BY A DESIGNATED INSTRUCTOR MAY OPERATE RADIO CONTROLLED AIRCRAFT AT THIS FIELD

INDEX

A
A.M.A. Membership 79
Aileron - Elevator Control 16
AILERONS 13, 16, 28
    Control 22
    Servo 46
    Trim 18
    Trim Control 22
Air Currents 26
Alcoholic Beverages 79
Altitude 29, 32
ALTITUDE and SAFETY 68
Antenna 14, 52
Approach, Gliding 76
Attitude 33
Attitude Test - Post-Start 56

B
Banana plug 39
Basic Maneuvers 74
Batteries
    Charging 12
    Chargers 42
    Condition 50
    Packs 12
    Trickle Charge Adapter 43
    12 Volt Motorcycle Type 40
Bolts - Nylon, Wing Securing 47
Breaking In the Engine 52
Buddy Cord 70
Bulb Fuel Pumps 41

C
Carburetor 19
    Check - After Engine Start 55
    Throat 18, 21
Carrier Wave 20
Channel Identification 45
Charge Connector 12
Charging Jack 15
Clevises 13
Climbing 28
Clunker 19
Combined Control - Aileron and Elevator 23
Communicating With Your Instructor 66
Construction 14

Control
  Horn 13
  Lever 15, 16
  Thumb only or Thumb and Forefinger 19
  Panel 37
  Sticks 15
  Surfaces 16
Courtesy 56
Covering Film 14

D
Descending 29
Digital Proportional Radio Control 18
Directional Control 28
Distance - Keeping Your Plane in Sight 69
Drugs 79
Duration of Flight 77

E
Electrically Powered Fuel Pumps 40
Elevator 16, 67
  Control 21
  Trim 17
  Trim Control 22
Emergency Transfer of Control 72
Empenage 14
ENGINE 16, 17
  Breakin 52
  Mounting 19
  Speed 30
  Starting 52, 81

F
Feathers 14
Feel of the Controls 73
Field Box 37
Field Box Glow Plug Heater 41
Field Equipment 37
Field Pattern 57
Fin 14
Firewall 19
Flaring at Final Touchdown 76
Flight Line 46
Flight Plan 73
Flight Procedures 80
Flying Field 35
Foam 19
Fuel
  Can 38
  Container 38

Fuel (Continued)
   Pump 37, 40
   Tank 16, 19, 49
Fueling 47
Full Scale Compared to Model Flying 31
Fuselage 14

G
Getting Acquainted with Your Model 11
Getting Ready to Fly 34
Glider 26
Glow Plug 17
   Heater 37, 41
   Heater - Connecting 54
Ground Maneuvering - Taxiing 58

H
Hand Operated Fuel Pumps 41

I
Instructor 66
INTRODUCTION iiii

L
Landing 75
   Deadstick 82
   Normal 82
Landing Gear 11
Landings 65
Left Aileron 67
Lift 25
Light Touch 19

M
Maiden Flight 56
Maneuvers - Basic 74

N
Nose wheel 11, 16

O
ON-OFF Radio Switches 14
Operating the Controls 20
Orientation 32
Overcharging 15

P
Performance 57
Pre-Flight Training 11
Preflighting Your Plane 35

Plane Size and Color 32
Plastic foam 12
Plastic wrapping film 12
Plug, Glow 17
Pocket Glow Plug Heater 42
Primary Flight Training 65
Priming 53
Propeller 16, 10

R
Rules for Safe Flying (Typical Club) 79
Radio
  Activation 50
  Channel Designation 80
  Channel Designator 81
  Channel - Clearance to Use 58
  Channel Identification 45
  Channels 44
  Controlled TRAINER Model Airplane 11
  Frequencies 44
  Frequency Control Board 80
  Prodecures at the Flying Field 44
  Range Check 11, 52, 81
  Receiver 12
  Transmitter 12
  Transmitter Impounding 46
  Transmitters at the Flying Field 45
Ribs 11
Right Aileron 67
Rubber bands 11
Rudder 11, 16, 28
Rudder - Nose Gear Connection 21
Rudder and Nose Wheel Trim 17
Rudder Trim Control 21
Rules of Conduct at Flying Fields 56
Rules for Safe Flying 79

S
Safety 24
Safety Coordinator 80
Sailplane 26
Seat of the Pants Flying 31
Servo 11, 13, 19, 20
Servo arm travel 18
Servo Reversing Switches 18
Shutting Down the Radios 23
Slide Controls 17
Spars 13
Speed Variation 55
Stabilizer 14

Stalls 29
Starter 39
Starter - Using 54
Starting the Engine 52
Steering 11, 60
Sticks 19
Structural Check 47
Switch 12
SYSTEM 12

T
Table of Contents iv
Tachometers 144
Tail Assembly 14
Takeoff 74, 81
Takeoff and Landing Direction 80
Takeoff Procedure 81
Takeoffs 65
Taking Over the Controls for the First Time 72
Taxiing 58, 81
Telescoping Antenna 15
Thermals 27
Third Dimension - Moving UP and DOWN 28
Throttle - UP, HALF, DOWN 68
Throttle Control 16, 18, 19, 20
Throttle Practice 59
Throttle Trim 17
Throttle Trim Control 20
Tools 42
Touchdown 76
TRAINING
   Safety Rules 82
   Solo Flight 83
   Transferring Control 69
Transmitter, Radio 14, 20
Trickle Charging 42
Trim 17
Trim Controls in Flight 57
Turning 28
Turns 67

V
Vision 69
Visual Observation 31
Voltmeter 43

W
Watching Other People Fly 63
What Makes Airplanes Fly ? 24

Wind 26
Wing 11
  Attachment 11
  Mounting 46
  Provides Lift 25
  Securing to Fuselage 47